Archaeoastronomy in the Old World

About this book
This volume summarises the proceedings of a conference which
took place at the University of Oxford in September 1981.
Held under the auspices of the International Astronomical
Union and the International Union for the History and
Philosophy of Science, the meeting reviewed recent research in
Old World Archaeoastronomy. The publisher received the final
typescript for production in March 1982. The papers on
archaeoastronomy in the Americas are also published by
Cambridge University Press in a companion volume,
Archaeoastronomy in the New World, edited by A.F. Aveni.

The papers in this book are concerned with shedding light on a
controversial aspect of European prehistory, especially that
of north-west Europe: was astronomy practised here in the late
neolithic and bronze ages, and, if so, what was its purpose?
These questions are of obvious interest to prehistorians, but
modern interest in them has been stimulated largely by those
whose professional background is in the pure and applied
sciences, while they raise technical issues which have aroused
the interest of statisticians and astronomers.

T0220440

Archaeoastronomy in the Old World

Edited by
D.C. Heggie

CAMBRIDGE UNIVERSITY PRESS
Cambridge
London New York New Rochelle
Melbourne Sydney

CAMBRIDGE UNIVERSITY PRESS
Cambridge, New York, Melbourne, Madrid, Cape Town, Singapore,
São Paulo, Delhi, Dubai, Tokyo

Cambridge University Press
The Edinburgh Building, Cambridge CB2 8RU, UK

Published in the United States of America by Cambridge University Press, New York

www.cambridge.org
Information on this title: www.cambridge.org/9780521125307

© Cambridge University Press 1982

First published 1982
This digitally printed version 2009

A catalogue record for this publication is available from the British Library

Library of Congress Catalogue Card Number: 82-4233

ISBN 978-0-521-24734-4 Hardback
ISBN 978-0-521-12530-7 Paperback

CONTENTS

PREFACE

The papers in this book are concerned with shedding light on a controversial aspect of European prehistory, especially that of north-west Europe: was astronomy practised here in the late neolithic and bronze ages, and, if so, what was its purpose? These questions are of obvious interest to prehistorians, but modern interest in them has been stimulated largely by those whose professional background is in the pure and applied sciences, while they raise technical issues which have aroused the interest of statisticians and astronomers. The diverse backgrounds of the authors of these papers reflect the multidisciplinary approach which the subject deserves and, indeed, requires.

The papers were presented at an international symposium on archaeoastronomy that was held at The Queen's College, Oxford, from 4 to 9 September, 1981. All the invited papers dealing with ancient astronomy in the old world are included, except for one introductory paper on the astronomical background, and the volume also contains a wide selection of the contributed papers, some of which were presented as posters. Papers dealing with American archaeoastronomy will be found in a companion volume edited by A.F. Aveni.

It is a pleasure to thank the authors of these papers for the time they have devoted to the preparation of their typescripts, and the staff of Cambridge University Press for their careful and expeditious production of this volume. This is also a suitable opportunity to thank the three bodies which materially supported the activities at the conference: the International Astronomical Union, the International Union for the History and Philosophy of Science, and the British Academy. But particular thanks are due to Michael Hoskin, who not only originally proposed the holding of the conference, but devoted much energy and good humour to its efficient organisation.

Edinburgh   Douglas C. Heggie

5 March 1982

CONTRIBUTORS

P.N. Appleton: University of Manchester, Nuffield Radio Astronomy Laborat-
        ories, Jodrell Bank, Macclesfield, Cheshire SK11 9DL, U.K.
R.J.C. Atkinson: Department of Archaeology, University College, P.O. Box
        78, Cardiff CF1 1XL, U.K.
H.A.W. Burl: 40 St. James Road, Edgbaston, Birmingham B15 1JR, U.K.
J. Čierny: Ruhr-Universität, D-4630 Bochum, West Germany
R.W. Few: 147 Girton Road, Girton, Cambridge CB3 0PQ, U.K.
P.R. Freeman: Department of Mathematics, The University, Leicester LE1 7RH,
        U.K.
A.J. Hastie: Cardonald College, 690 Mosspark Drive, Glasgow G52, U.K.
D.C. Heggie: Department of Mathematics, University of Edinburgh, King's
        Buildings, Mayfield Road, Edinburgh EH9 3JZ, U.K.
A.Lynch: National Monuments, Office of Public Works, 51 St. Stephen's Green,
        Dublin 2, Ireland
T. McCreery: Cardonald College, 690 Mosspark Drive, Glasgow G52, U.K.
E.W. MacKie: Hunterian Museum, The University, Glasgow G12 8QQ, U.K.
T. Moulds: 9 Park Road, Glasgow, U.K.
R.P. Norris: University of Manchester, Nuffield Radio Astronomy Laborator-
        ies, Jodrell Bank, Macclesfield, Cheshire SK11 9DL, U.K.
J.D. Patrick: Department of Computer Science, Deakin University, Victoria
        3217, Australia
O. Pedersen: History of Science Department, University of Aarhus, Ny Munke-
        gade, DK-8000 Aarhus C, Denmark
K. Piasecki: Zakład Antropologii Historycznej UW, Krakowskie Przedmieście 1,
        00-927 Warszawa, Poland
G.H. Ponting: 'Olcote', Callanish, Isle of Lewis, PA86 9DZ, U.K.
M.R. Ponting: 'Olcote', Callanish, Isle of Lewis, PA86 9DZ, U.K.
J.N.G Ritchie: The Royal Commission on the Ancient and Historical Monuments
        of Scotland, 54 Melville Street, Edinburgh EH3 7HF, U.K.
C.L.N. Ruggles: Department of Archaeology, University College, P.O. Box 78,
        Cardiff CF1 1XL, U.K.
R.M. Sadowski: ul. Międzynarodowa 33 m. 24, 03-962 Warszawa, Poland
W. Schlosser: Ruhr-Universität Bochum, Astronomisches Institut, Postfach
        102148, D-4630 Bochum 1, West Germany
A. Thom: The Hill, Dunlop, Kilmarnock KA3 4DH, U.K.
A.S. Thom: The Hill, Dunlop, Kilmarnock KA3 4DH, U.K.
C.S. Wallace: Department of Computer Science, Monash University, Clayton,
        Victoria 3168, Australia
M.S. Ziółkowski: Zakład Antropologii Historycznej UW, Krakowskie Przedmieście
        1, 00-927 Warszawa, Poland

MEGALITHIC ASTRONOMY: HIGHLIGHTS AND PROBLEMS

D.C. Heggie
3 St Ninian's Terrace, Edinburgh EH10 5NL, U.K.

Abstract. After a discussion of the need for statistical methods in megalithic astronomy, the interpretation of statistical results is considered, and a simple method of performing a suitable statistical test is outlined. Statistical evidence is considered next, with particular regard to the extent and limitations of the role of selection effects in the work of the Thoms. Their most recent analysis of lunar lines is discussed. Some arguments of a practical nature then follow, and the paper ends with some briefer comments on the astronomical interpretation of megalithic art, dating, and the purpose and implications of megalithic astronomy.

INTRODUCTION

The present paper is a rewritten version of a review article[1] (Heggie 1981b) which was designed as an introduction to problems of megalithic astronomy for participants at the conference in Oxford. However the resemblance is largely confined to the title and the structure. Its general purpose here is to provide a background for many of the other papers in the present volume, but the opportunity is also taken to argue the case for the use of a particular methodology in the examination of much of the evidence on megalithic astronomy, and to consider how it may be applied to one or two important bodies of data. Another purpose of this paper is to draw the attention of readers to a number of recent results which have a bearing on the subject and are perhaps not mentioned elsewhere in this volume. On the other hand nothing is said of the archaeological background to these studies, which is one of the purposes of the paper by Dr Ritchie, and little space is devoted here to the numerous important archaeological arguments which bear on the interpretation of the evidence; but many of these are discussed by several of the authors who have contributed to this book.

The idea that astronomy was one element in the function of certain megalithic sites in Britain and elsewhere is an old one, as Prof.

Atkinson mentions in his paper, but it has never flourished more vigorously
than in the last two or three decades, thanks largely to the work of Prof.
Alexander Thom. Though his work had begun in the 1930s, it was not until
1954 that his first paper on the subject of prehistoric astronomy appeared.
In 1967 he published a book (Megalithic Sites in Britain) in which evidence
from some hundreds of sites was collated (Fig.1). Much new evidence was
then produced in two subsequent books (Thom 1971; Thom & Thom 1978a),
where particular emphasis was laid on very accurate sightlines for the
moon using distant markers, mostly natural.

       The present activity in the field of archaeoastronomy in the
Old World, or at least in the British context, can be seen largely as the
response of experts in different disciplines to Thom's work. Among
archaeologists some, such as Aubrey Burl, have selected the parts of Thom's
ideas which can be absorbed most readily into a fresh but relatively
conventional picture of the societies of the megalith builders. A few,
notably E.W. MacKie, have considered in a more radical way what changes in
the conventional picture would be required by Thom's theories. And it must
be admitted that there are many archaeologists, who do not attend conferences on archaeoastronomy, who consider that his theories shed no light on
these problems, for all the acknowledged excellence of his field surveys.
Finally there are those with less archaeological experience, or, more often,
none at all, who have responded actively to the more technical aspects of
Thom's work, such as the statistical evaluation of his theories, and the
associated problems of ensuring that the available data are suitable. But
Prof. Thom's work has not been limited to the astronomical aspects of
megalithic sites, and if the response to his work on megalithic geometry
has been less vigorous, it is because the necessary methods of analysing
his work in this field have proved much harder to develop. (Dr Patrick's
paper in this volume marks a significant breakthrough in this respect.)
Nor is Thom's influence confined to Britain, as many of the American
archaeoastronomers at the Oxford symposium testified warmly to the great
interest which his work has engendered in the New World also.

STATISTICAL ARGUMENTS
The need for statistical methods

       The great bulk of the evidence on megalithic astronomy consists
of orientations, alignments, or 'lines', directed towards a place on the
horizon where a conspicuous astronomical object (sun, moon, a planet, or a

bright star) rises or sets. Of course the rising and setting positions of
the sun, moon and planets vary relatively rapidly, and so for them it is
the extreme rising and setting positions that are generally considered.
For the sun the extreme positions are those reached at the solstices, but
several authors have also considered orientations for the sun at other

Fig.1. The astronomical sites listed in Thom 1967.

times of the year, as discussed, for example, by the Thoms in their paper in this volume. The rapid and relatively complicated motion of the moon makes its extremes rather involved, but at a low level of accuracy the extreme positions are simply the 'standstill' positions, to use the convenient term introduced by Thom (1971, p.18). Extreme positions for the planets may also be defined, but in fact the planets have been rarely discussed in this context.

Coupled with this variety of astronomically significant positions is an equally broad choice of megalithic orientations. These may be defined by single slabs, true alignments of standing stones, lines from the centre of a stone circle to an outlying stone, lines from one site to another, the axes of megalithic tombs, lines from a prehistoric site to a natural horizon feature (or 'foresight'), such as a hill-slope, a valley or notch, and so on.

Given this great diversity of 'targets' and orientations, one must consider the possibility that it is only by chance that one finds some sites to be astronomically orientated. Indeed we can be virtually certain that coincidences occur, and since we do not know a priori that the megalith-builders were interested in any particular astronomical object, it seems wise to dismiss any apparently astronomical orientations as coincidences unless there is evidence to the contrary. If we relax this attitude (and it must be said that many authors have never adopted it in the first place) then we shall be in grave danger of writing books and papers whose rightful place is on the fiction shelves. The fact that it has been ignored so often possibly accounts for the general air of controversy in which debates on megalithic astronomy tend to have been conducted. If indeed a body of evidence does not allow us to conclude that the orientations involved are not just coincidences, it is perhaps to be expected that an enthusiastic archaeoastronomer and a careful archaeologist can come to diametrically opposed conclusions.

The stress which is laid on statistical methods here (and in Prof. Freeman's paper in this volume) reflects the nature of the evidence available for the study of megalithic astronomy. One can imagine plausible circumstances in which the importance of such methods might be much diminished. In his paper in this volume Dr MacKie expresses the hope that a site might be found which, on excavation, decisively supports only an astronomical interpretation. Were such a site to be found then statistical evidence would have only a small role to play with regard to that site and others

like it. But at present no such key site is known, and the great bulk of
the evidence requires statistical evaluation if we are not to be misled by
it. The study of Mayan astronomy, on the other hand, exemplifies a field
where the statistical evaluation of orientations is of relatively minor
importance, because information from orientations is supported by evidence
of the most decisive kind, in the form of a decipherable astronomical
notation. At the Oxford conference, several European participants voiced
some dismay at the lack of statistical analysis of the evidence on native
American astronomy, but it may be argued that this reflects the rather
supplementary role which the study of orientations plays in much of this
work. The great difference in the nature of the evidence available on
the two sides of the Atlantic is one of the points to which Prof. Pedersen
also draws attention in his paper in this volume.

In the context of megalithic astronomy, other evidence of a
decisive kind (i.e. apart from orientations) has not yet turned up, and
orientations are the basic evidence on which the hypotheses have to be
tested. And to establish that the orientations one finds are not coincidences one has to show, if possible, that one finds more orientations towards phenomena of astronomical importance than would be expected to occur
by chance. The most important role of statistics in this context is, in
effect, to calculate the probability of obtaining by chance a number of
orientations not less than the number actually found in the investigation
under study. If the orientations have nothing to do with astronomy and
are essentially random, then the probability thus found is unlikely to
take very low values. Sometimes, as with the statistical test devised by
Freeman & Elmore (1979), the results are not expressed directly in terms
of probabilities, but they can still be interpreted qualitatively in much
the same way.

Interpretation of statistical results

Before considering how the probability is to be calculated, we
must discuss how the results are to be interpreted. If the probability
obtained is not very low, then we have little reason to reject the notion
that the orientations are random. (This is not to say that the orientations
are not deliberate, however, for one might contemplate the suggestion that
the sites were orientated on non-astronomical objects, such as hills, or
sacred places, or routeways. The point is that, even if such a suggestion
were correct, we would have no reason to expect more orientations with

apparent astronomical significance than we would expect from a genuinely
random set.) If, on the other hand, the probability turns out to be very
small, then we have strong grounds for rejecting the notion that the
orientations are random, but we must take care before coming to the obvious
conclusion that they are genuinely astronomical. Low levels will occur
only rarely if the orientations are random, but they will occur. In the
same way, if we look through enough sets of randomly generated data, we
will be able to select one or more which look very non-random. This is the
lesson which the rather outré example in Dr MacKie's paper in this volume
ought to teach us. Unless our data have been selected objectively, the
obvious conclusions may easily be false ones, and Aubrey Burl's recommend-
ation (see his paper in this volume) that one should study groups of
monuments, rather than individual examples selected from such groups, can
be seen as a safeguard against this.

The most important selection effects that can endanger the
statistical analysis of megalithic orientations are concerned with the
selection of the sightlines themselves. It has been realised for a long
time (Thom 1955) that it is difficult to be objective about this, and it
is no less difficult to try to apply strict selection criteria retrospect-
ively. Nevertheless this must be attempted if we are to derive meaningful
conclusions from most of the existing evidence for megalithic astronomy,
at least until we possess fresh bodies of data prepared according to strict-
er selection criteria. (In his paper in this volume Dr Ruggles offers us
precisely this prospect.)

This emphasis on selection criteria should not obscure the other
problems associated with the statistical investigation of orientations,
though they are often more easily overcome. For example, one must try to
minimise the influence exerted by the data on the formulation of the particular astronomical hypothesis to be examined.

A simple statistical method

Finally we come to the way in which the probability is to be
calculated, i.e. the probability that a set of random orientations would
yield (by chance) the number of astronomically significant lines that we
actually find, or even more. A simple and rough (but widely applicable)
method of doing so will now be described.

Several pieces of information are required, namely
(i) the tolerance, t, i.e. how closely an orientation must agree with an

astronomically significant direction to be included. The tolerance may
depend on the accuracy with which the direction of the orientation can be
defined, in which case it will generally be expressed in terms of azimuth.
But it may also be expressed in terms of declination, and then it can be
converted approximately into an equivalent tolerance in azimuth by multi-
plying by the quantity $|dA/d\delta|$ (the rate of change of azimuth with declin-
ation). For many approximate purposes this may be read from Fig.2. As an
example, an orientation to some point on the sun's disc at a solstice must
indicate a declination within approximately 16' of that of the sun's centre,
i.e. the tolerance in declination is 16', since this is the mean apparent
radius of the disc. Since the declination is about $24°$, at the latitude of
Stonehenge ($51°.2$) $|dA/d\delta|$ is about 1.9, leading to a tolerance in azimuth
of about 30'.

Fig.2. The factor $|dA/d\delta|$. The graph is entered by declination
(δ) and latitude (λ). The curves were calculated for the case
of zero horizon altitude, and the sign of the declination is
irrelevant. It is not possible to obtain accurate values from
these graphs at declinations close to the colatitude ($90°-\lambda$).

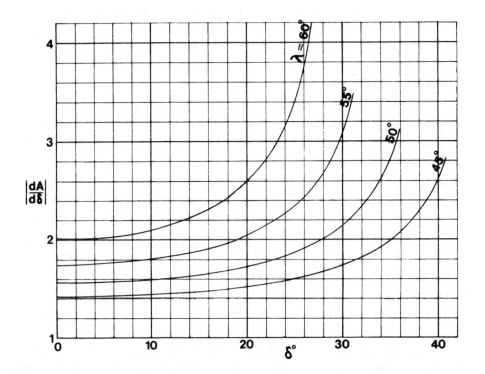

(ii) the number of astronomically significant positions. These must be
chosen fairly (Hawkins 1968), so that if the position of sunrise at the
equinox is included, then so must the position at sunset. Again, if we
wish to test the hypothesis of the solstitial orientation of the axis of
Stonehenge, we must include both sunrise and sunset at both solstices, i.e.
four positions in all.

(iii) the probability, p, that a single random orientation would be regarded as astronomically significant. In straightforward cases this is simply
the number of significant positions times twice the tolerance in azimuth
(since an orientation may deviate on either side of the precise astronomical
direction under consideration), divided by 360°. If we are testing the
solstitial hypothesis for the orientation of the axis of Stonehenge, then
$p = 4 \times 2 \times 30'/360° = 0.011$, approximately.

(iv) the total number of orientations, N, in the sample, i.e. including
both astronomical and non-astronomical orientations. What is actually needed is the total number of available orientations from which the sample was
selected, and it is often difficult to find information on this. Indeed it
is at this point that it is frequently necessary to make some allowance for
unwritten or faulty selection criteria. In the case of the axis of Stonehenge, however, it is clear that one should take N=1, since the axis is
much the most clearly defined orientation at Stonehenge, and may be regarded
as the unique principal orientation of the monument (cf. Prof. Atkinson's
paper in this volume).

(v) the actual number, n, of astronomical orientations in the sample. For
example, since the axis of Stonehenge does actually indicate a point on the
sun's disc at the position of midsummer sunrise at the time of its construction (this can be inferred from the data given by Thom & Thom 1978a, p.150),
we have n=1.

          Now we have to calculate the probability, P, that at least n
lines out of a sample of N random orientations will indicate some astronomical position within the stated tolerance. If n and N are both 1,
as in the test of the axial orientation of Stonehenge, then obviously P=p.
In other cases, P may be calculated from expressions based on the binomial
distribution (the standard formula is given, for example, in Heggie 1981a,
p.242), or from an approximation based on the Poisson distribution.
For the latter purpose we need also

(vi) the expected number, n_e, of astronomical orientations, which is just
the average number that would be expected to occur in a sample of N random

Fig.3. Probability, P, of obtaining at least n astronomical
orientations by chance. The average number that would be expected by chance is denoted by n_e. Above the line P = 0.001, the
value of P is below 0.001; below the line P = 0.1, the evidence for astronomical orientations is not statistically significant. For restrictions on the applicability of this graph see
the text.

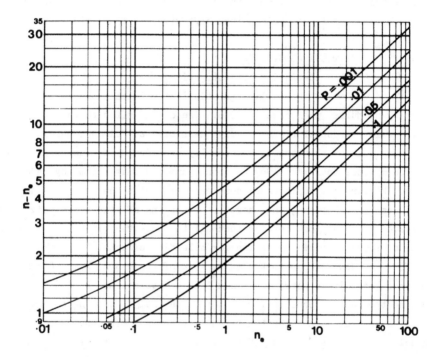

orientations, i.e. n_e = Np. If N is large and n_e is small then the
Poisson approximation is satisfactory, and then the probability P can be
read from Fig.3, which is entered by the expected number n_e and the excess
of the actual number over the expected number, i.e. $(n-n_e)$. Actually, the
probabilities obtained from this graph are also approximately correct in
other circumstances, e.g. when n = 1 and n_e is very small, but without
any restriction on N. Indeed the correct result for the Stonehenge problem
$(n_e = 0.011, n = 1)$ can be obtained approximately from this graph.

STATISTICAL EVIDENCE
Early statistical investigations
        Some of the earliest statistical arguments are to be found in a
book by R. Müller published in 1936. One of the sites to which they were
applied is a group of rings at Odry (near Czersk, Poland), but in fact these
are much younger than the sites normally considered in studies of megalithic
astronomy (see Dobrzycki 1963, and the paper in this volume by Sadowski et
al.). Müller also rediscussed some earlier statistical results on a group
of lines by J. Hopmann, and showed that nothing of statistical significance
survived when one took into account all the possible sightlines - an early
case of retrospective correction for selection effects.
        Now we come to the early work of Professor Thom. His first
paper on megalithic astronomy (Thom 1954) contained no statistical analysis
of the type we are considering, but such an analysis, using the data from
this paper, has been attempted recently (Heggie 1981a, pp.153f.), and the
conclusions are as follows. Formally there is statistically significant
evidence for solstitial orientations (of three distinct structural types,
but analysed together) in Thom's paper. This is true even if we are rather
strict in omitting sightlines over which, in Thom's opinion as expressed
here and in later books and papers, some doubt exists. Furthermore Thom's
own selection criteria in this paper seem to have been quite strict, to the
extent, incidentally, of excluding sightlines (such as that to the south-
west at Ballochroy, Argyll) which have been brought subsequently to some
prominence in later papers and which, if retained, would have strengthened
the evidence for megalithic astronomy. These solstitial sightlines also
strongly suggest that it was not just any point of the solar disc, but one
or other limb, towards which they were directed, though no probability level
was calculated for this result.
        Professor Thom's second paper dealing with megalithic astronomy

appeared in 1955, and contained (for the first time in his work) an analysis
of the statistical significance of the evidence, in relation to solstitial
and equinoctial sightlines taken together. What was also new about this
paper, however, was the discussion of possible stellar lines. Now since
there are so many stars, and since their rising and setting positions vary
by such large amounts over the period in which the megalithic sites were
built, several writers have been apt to dismiss any talk of stellar orientations out of hand on the grounds that these can so easily be coincidences.
The fact that such comments are rarely challenged demonstrates that this
early paper of Thom's has not received the attention that it certainly deserves, for in it he shows that the number of stellar orientations which he
found greatly exceeds what would be expected in a comparable sample of random orientations. This excess occurred only for a narrow range of dates -
about 2100 B.C.

   As usual with such studies, it is wise to examine this investigation carefully, to ensure that the result is not obviated by selection
effects. In fact the sightlines were of two types, but Thom himself noted
a number of difficult, borderline cases, and a re-examination of the selection of the sites in this paper (Heggie 1981a, pp.157f.) raises similar
doubts over other sites which Thom did not specifically note as posing any
difficulty. However, as Dr Ruggles points out in his paper in this volume,
Thom did remark in his paper that 'nothing has been excluded which has a
bearing on the subject', and it may be that selection effects play a relatively insignificant role.

The evidence in Thom's first book

   The next major publication by Prof. Thom was the epochal 'Megalithic Sites in Britain' (1967). This book contains, besides numerous
comments on individual sightlines, a master list of over 250 lines ('Table
8.1') from almost 150 sites (Fig.1). It must be stressed that Thom himself
did not attempt to use this data for any quantitative estimates of statistical significance, but several later writers have quoted this list of sightlines and an accompanying histogram ('Fig.8.1') as powerful evidence for the
practice of megalithic astronomy. What has generally been neglected in such
assessments is any consideration of the nature of the criteria according to
which these lines were selected for inclusion. Thom himself recognised the
varying reliability of these, by assigning each line to one of three classes,
from class A (which includes the most reliable lines) to class C. The latter

includes lines which are admitted merely because they indicate a declination
of astronomical significance, and of course it would not be possible to use
these for a fair statistical test of the astronomical hypothesis, as Thom
mentions. Nevertheless questions about the selection criteria, which are
discussed in a general way by Dr Ruggles in his paper in this volume, are
not restricted to class C, but apply also to some lines of class A, as the
following examples suggest.

At Ballochroy (Argyll) one of the three stones indicates a natural foresight (in this case a hill-slope) for midsummer sunset, and this is
assigned to class A. However, one of the other stones also indicates a hill-
slope, but in this case no astronomical interpretation is known (MacKie 1974),
and this line is not included in Thom's list at any level of reliability.
Thom accounts for this (pers. comm.) by noting that the central stone is
the taller of the two. However it was MacKie's view that 'if one considers
the site without foreknowledge or preconceived ideas there is no reason
from its layout to prefer either ·f the indicated alinements as the primary
solstitial one', and so the line i iicated by the shorter stone should have
been included at least in class B, which 'contains borderline cases which
some people might accept and others discard' (Thom 1967, p.96).

Another example occurs at Castle Rigg (Cumberland), where two
of the four sightlines (all of which are of class A) are provided by one of
the special diameters, used in both directions, in Thom's geometrical construction for this ring. Both sightlines have an astronomical interpretation.
However the other special diameter (symmetrically disposed with respect to
the first) is not listed in either direction in any class. Its ends, like
those of the listed diameter, are marked by stones, one of these being one
of the tall entrance portals (Burl 1976, pp.58-9; this paper, Fig.4). Also,
the horizon altitudes (information on which can be found in Thom 1966, Fig.
39) are comparable, and the unlisted diameter is notable for its indication
of a very prominent foresight to the north. To the south it could perhaps
have served as an indication of the moon at the major standstill, but to the
north it could not have indicated any of the astronomical declinations considered by Thom in his book.

A third example which illustrates the selection difficulties in
the list in Thom's book is the little alignment at Doune (Perthshire), of
which photographs will be found in Heggie 1981a, p.161. Though details of
the alignment in both directions were quoted in an earlier paper (Thom 1955),
only the line to the north was listed in the book, where it was assigned to

class A. In this direction the line can be identified with the star Capella,
but to the south it indicates none of the astronomical declinations considered in the book.

Evidence on accurate lunar lines

Since the time of publication of his first book, Prof. Thom's
attention has been much taken up with the study of supposed very accurate
lunar lines. One of the first of these studies to have been tested statistically was his work on the lunar 'observatory' using Le Grand Menhir Brisé,
near Locmariaquer (Brittany), as a foresight. The result (Freeman 1975)
was that there was nothing of significance to be explained if the number
of potential backsights (i.e. possible observing positions) exceeded about

Fig.4. An unlisted sightline at Castle Rigg, viewed from the
north. The line, which is a diameter of Thom's geometrical
construction for this ring, runs from the east portal (foreground, right) to the middle stone of the group of three on the
far side of the ring. The other (symmetrically disposed) diameter is illustrated in Heggie 1981a, pp.112,113.

two hundred. It was the need to make retrospective allowance for possible
effects of selection of backsights that led to a conclusion of this form.
Actually doubts have been voiced over even those backsights that were
chosen (Hadingham 1981a), and over the foresight, since there is no record
that the Menhir ever stood upright (see Burl 1976, pp.129-30), though there
are certainly arguments for supposing that it did (Atkinson 1975; Merritt
& Thom 1980).

         The Thoms have identified many other possible lunar sightlines
of high accuracy, and have used these to test the theory that the purpose
of the lines was to record fine variations in the extreme declinations of
the moon caused by solar perturbations. If this theory is correct, one
expects the indicated declinations to be concentrated at particular values
within the so-called lunar 'bands' (which are defined by the Thoms in their
paper in this volume). If, on the other hand, the lines were at most relatively rough indications of the lunar extremes, and the perturbation was
not observed, then one might expect the indications to be relatively randomly distributed over the bands. In fact from several analyses (Thom & Thom
1978a, p.136; 1978b; A.S. Thom 1981) the Thoms have concluded that the
latter hypothesis must be rejected, and that the indicated declinations are
concentrated, at a high level of statistical significance, towards the values expected on the hypothesis which they proposed. But they have also
expressed surprise that the indicated declinations lie as close to the expected values as they do, because the agreement seems closer than one would
think possible on practical and other grounds (see also Heggie 1981b).

         It may be that an explanation of this difficulty is to be sought
in selection effects. When some attempt is made to allow for other potential combinations of backsights and foresights at a few much-studied sites -
such as Temple Wood in Argyll and Brogar in Orkney (see Heggie 1981a, pp.
172f.) - it appears that the number of indicated declinations close to the
expected values is not so different from what would be expected if the indicated declinations were randomly spread over the lunar bands. A thorough
examination of selection effects has yet to be completed (Ruggles 1983),
but the foregoing examples are surely indicative. Furthermore the analysis
made by the Thoms is quite sensitive to the selection of the foresights.
For example, their formally most significant result (A.S. Thom 1981) concerned 42 sightlines, and for this sample the probability of obtaining such
close agreement with the ideal declinations is about 10^{-4}. But if one added
only another dozen or so foresights indicating declinations within the lunar

bands, but far from any of the ideal declinations, it would no longer be possible to conclude that the indicated declinations were non-random, at any level of significance. It cannot be argued that the selection of fore-sights is complete to anything like this extent.

The Thoms' new analysis

It is the problem of the influence of selection effects in previous analyses which makes the new analysis reported by the Thoms in their paper in this volume so important. Here sightlines are divided into three classes, depending on the sign of Δ (the solar perturbation) in the expression for the declination, δ, of the centre of the moon in terms of ε (the obliquity of the ecliptic), i (the mean inclination of the lunar orbit), and Δ. (A lucid discussion of the meaning of these quantities will be found in Morrison 1980.) Lines for which $\delta = \pm \varepsilon \pm (i+\Delta)$ are called 'favourable' (F), those where $\delta = \pm\varepsilon \pm (i-\Delta)$ are called 'unfavourable' (U), and the line is not relevant if $\delta = \pm\varepsilon \pm i$. The important point is that the con-cept of favourable and unfavourable lines was formulated after the select-ion of the foresights was made, and so could not have influenced the sel-ection. Hence, the argument runs, if the foresights were distributed at random there should be on average equal numbers of U and F declinations, whereas the actual numbers are significantly different.

A couple of detailed remarks may be made about this. First, when one examines the ideal declinations listed by the Thoms (the paper in this volume, Table 1), the range of declinations corresponding to U sight-lines is found to exceed slightly the range corresponding to F sightlines, which makes their result still more significant. Secondly, however, when one talks of a random distribution of foresights in the lunar band, one is thinking of foresights distributed at random in <u>azimuth</u>. The distribution of indicated <u>declinations</u> is therefore proportional to the factor $|dA/d\delta|$ plotted in Fig.2. Now this factor increases as the arithmetic size of the declination increases. Since the favourable and unfavourable declinations are different, a uniform distribution of azimuths gives rise to different distributions of declinations near the favourable and unfavourable values. The effect is negligible for the minor standstill, but is more significant for the major standstill, especially at high latitudes when the foresight is at a high altitude in the southern half of the horizon. At a very few sites the effect is so strong that the probability of obtaining a favourable declination may exceed about 0.6. Nevertheless such a case is exceptional,

and it has been estimated that the average value of the probability is about
0.53 (when account is taken of both effects referred to in this paragraph).
This implies that the expected number of favourable declinations in the
Thoms' sample of 51 lines increases from 25.5 to about 27.0. This change
is large enough to alter their probability level by a very substantial factor (which can be estimated from Fig.3), but not by so much as to convince
one that the result can be explained away.

It may well be, then, that we here have significant evidence
for an excess of favourable over unfavourable foresights. But the inference to be drawn from this may well only be that, while the unfavourable
foresights are coincidences, a few of the favourable ones are intentional.
If so, then what we may infer is only that there are deliberate foresights
for the moon when the declination of its centre is $\delta = \pm\epsilon \pm(i+\Delta)$; and,
although this expression contains the perturbation term Δ, we are not able
to infer that the variation caused by the perturbation was observed, but
only the extreme declination, which occurs when the perturbation is at maximum. The possibility of such a conclusion (on other grounds) was also
considered by Morrison (1980), and it could remove the need for methods of
extrapolation (Thom 1971, ch.8) and associated problems. But while our
tentative conclusion may seem a rather modest one, it could be quite important. Previously it was thought by many that the best evidence for
sightlines to the extreme lunar declinations was that contained in Thom
(1967,chapter 10), but this evidence no longer seems very sound (Ruggles
1981). Therefore, if, as is suggested here, the new discussion of favourable and unfavourable foresights lends some support to the same conclusion,
then this discussion may be of some importance.

One may ask, with all the attention devoted to lunar lines in
the last decade, what has happened during that period to Thom's earlier
astronomical theories? As far as solar or calendrical sightlines are concerned, the answer will be found in the Thoms' paper in this volume. But
stellar lines have been a casualty of their revisions over the years. One
particularly conspicuous group of stellar lines in the 1955 paper was re-
interpreted as being calendrical (Thom 1967, p.103), and some other stellar
lines were reassigned to the moon (those discussed in Thom & Thom 1978a,
p.5, and one or two others discussed in Thom 1955 which might at that time
have been identified with the star ϵ Canis Majoris, or Adhara). If the
discussion of possible selection effects in the later work has the effect
of diminishing the support for the lunar theory, then the time may come for

a reexamination of these earlier interpretations. Certainly they were
based on a sample of sightlines whose selection has not yet been called
into such serious question, and their statistical significance is relatively clear.

Other statistical research

There are several statistical investigations which have not
been mentioned here. But the statistical evidence on Stonehenge has been
summarised recently (Heggie 1981a, pp.145f., pp.196f.), and the possible
astronomical aspects of this monument are the subject of Prof. Atkinson's
paper in this volume. Much of Aubrey Burl's recent work, which can be
traced through his paper in this volume, is susceptible to statistical analysis, where conclusions are not obvious from a visual inspection of the
data. Several other papers in this volume also contain evidence of a statistical nature, including those by Freeman, Norris et al., and Lynch.

Particular mention must also be made here of some other recent
statistical investigations of prehistoric astronomy in Europe, those of
grave orientations in cemeteries. Though not concerned with megalithic
monuments, these studies certainly rely on evidence which, in its nature
and its analysis, closely resembles that for megalithic astronomy, and two
papers at the Oxford symposium were devoted to this material. An account
of the work of K. Barlai on Hungarian sites is already in print (Barlai
1980), and the present volume contains a paper which summarises results by
W. Schlosser and J. Čierny on some different sites in central Europe.

NON-STATISTICAL ARGUMENTS

The statistical method offers a rational way of deciding how
well the distribution of orientations agrees with some variant of the
astronomical hypothesis; and since most of the influential work on megalithic astronomy rests on the presentation of data showing that orientations
appear to agree with astronomically significant directions, the statistical
examination of these results is bound to play an important role. But there
are several other yardsticks against which the hypotheses can be measured,
such as feasibility, and some types of evidence can have a bearing on the
subject in a way that is not amenable to statistical analysis. It is the
purpose of this section to consider a few of these arguments.

Practical arguments

Questions of the visibility of suggested foresights at certain
sites have led to some interesting results. One concerns a supposed lunar
line to the south at Callanish 1, in the Outer Hebrides. The line in question consists of an accurate foresight for the moon (Thom 1967, pp.124-5)
which is indicated by the avenue at Callanish 1. But it has been known for
some time (Hawkins 1971) that the view of the foresight from the avenue is
obscured by an outcrop of rock to the south of the site. (For illustrations
see Burl 1976, p.151, and Roy 1980.) This led Hawkins to suggest (Hawkins
1973, p.240) that the intention was to make the moon appear to set into
this rock, and the Pontings have suggested that the site might have been
arranged deliberately so that the moon set within the stones of Callanish 1
(see Fig.2 of their paper in this volume), an idea that parallels Burl's
suggestions about the possible lunar orientation of certain recumbent-stone
circles (Burl & Piper 1979, p.36). The presence of the obscuring outcrop
tends to detract from the idea that the foresight might have been used for
precise observation of the moon, a possibility that is weakened further by
an argument about extrapolation (Heggie 1981a, p.195). This problem of the
visibility of the foresight at Callanish incidentally illustrates something
which is rather typical of such structural arguments: rather than allowing
us to conclude that astronomy was not practised at a site, an argument of
this kind is usually more of a clue to the nature and purpose of any astronomy practised there.

But there is one site at which the problem of the visibility of
a foresight led to discoveries of a quite unexpected and intriguing nature.
At Kintraw, Argyll, the visibility of the solstitial foresight from the
backsight was found to be inadequate because of an intervening ridge (Thom
1967, p.155), and so it was suggested (Thom 1969) that observations were
made from a ledge in the hillside behind the site. E.W. MacKie then made
the remarkable discovery by excavation of a layer of stones on this ledge
(MacKie 1974), but acceptance of the implications of this work has been
hindered by two things. One is the absence of clear evidence, in the form
of artefacts, for instance, that the layer of stones is artificial, but the
other was a persistent problem over the visibility of the foresight from the
ledge. MacKie found that the mountain which forms one side of the foresight
'can be seen clearly over the ridge (in good weather)', and published a
photograph illustrating the fact (MacKie 1977, p.106). J.D. Patrick, on the
other hand, 'could not satisfy himself that it was possible to see the base

of the notch using a theodolite with 30 power magnification', a statement
again accompanied by an illustrative photograph (Patrick 1981). Now the
paper in this volume by McCreery et al. goes a long way towards resolving
this disagreement: it seems that both authors are right, but only some of
the time.

Another piece of evidence based on practical considerations,
and one that is not amenable to any statistical calculation, is the siting
of two of the mounds at the Ring of Brogar, Orkney (see the paper by the
Thoms in this volume). On the other hand, one practical argument which has
tended to be forgotten in recent years is the need for adequate space for
extrapolation, at sites which are considered to be accurate lunar observatories (Thom 1971, chapters 8,9). For the foresight to the north-north-
west at Brogar, i.e. Ravie Hill, the extrapolation length 4G (in Thom's notation) is about 0.8 km. If, therefore, observations were made by moving at
right angles to the line of sight, some observations must have been taken
from boats on the Loch of Harray - a most novel form of megalithic backsight!

Astronomical interpretation of megalithic art

These practical arguments are concerned with orientations and
associated questions, and the bulk of the evidence on megalithic astronomy
is of this nature. But there are independent pieces of evidence for which
statistical methods are of no avail. An example is what has been called
the 'Calendar Stone' at Knowth (Co. Meath), which has been described recently (amid much nonsense, it must be said) by Brennan (1980, p.98). On the
upper half of this stone (Fig.5), along with other markings, is a roughly
oval sequence of 29 symbols. The seven uppermost symbols are either a
circle or a pair of nested circles, and the remainder, which mostly extend
in a horizontal row along the middle of the stone, are crescents. This
design can be interpreted as a cycle of symbols representing the motion of
the moon through one synodic month (new moon to new moon) of 29.5 days, with
a rough representation of the phases and of their typical relative prominence.
Although the designs and markings on some other megaliths have often been
interpreted in astronomical terms, no example is as persuasive as this.
One case which has been shown recently to be false is Müller's interpretation (Müller 1970, p.107) of the markings on an orthostat in the tomb called
'Table des Marchands', close to the Grand Menhir. According to Müller there
are 56 individual markings, which represent the years of three nodal cycles
of the moon, as with the 56 Aubrey Holes at Stonehenge (Hawkins & White

1965, pp.177f.[2]), but it has been pointed out (Hadingham 1981b) that there
are, quite simply, fewer than 56 symbols on the stone, and perhaps only 49.

MEGALITHIC ASTRONOMY IN THE WIDER CONTEXT

The study of megalithic astronomy is part of the study of megalithic monuments in general, and this is the business of prehistorians and
archaeologists. Therefore everyone involved must be concerned to ensure
that sound results on megalithic astronomy can be properly encompassed within the larger field of enquiry. The observation, often made, that non-
archaeologists researching into the subject ought to acquaint themselves
better with what archaeologists have found out, is one to which all would
give their assent, but there are broader aspects of archaeological thinking
(for example, on the nature of the societies in question, or on the relationships between monuments of different types) of which the typical non-
archaeologist's view remains ill-focussed and fluid. Here it is difficult
for the non-archaeologist, and perhaps unwise, to do more than pass the odd
comment, and to ensure that the discussion is based on sound facts and
straightforward interpretations, to the extent that this lies within his
competence.

One obvious area of contact is dating. Commenting on the previous version of this paper, Dr Ritchie kindly pointed out that it was wrong
to imply that anything could be inferred about the date of Brogar from the

Fig.5. A possible lunar representation on the 'Calendar Stone',
Knowth (schematic, after Brennan, 1980).

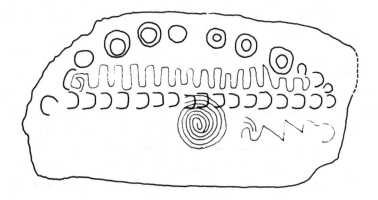

radiocarbon dates obtained for the nearby monument at Stenness, and so there was no conflict with the narrow range of dates for Brogar inferred by the Thoms on astronomical grounds. Actually the Thoms also regard Stenness itself as a lunar backsight (Thom & Thom 1978a, p.130) at a date close to those assigned for the lines at Brogar, i.e. within about 150 years of 1460 B.C., some 1500 years after the dates for Stenness. But strictly there is no contradiction, since the foresight for Stenness is a cairn, and it may have been this rather than the backsight which was positioned to define the sightline.

If the lunar line at Stenness is not simply an accident, it exemplifies one of the problems confronting the archaeologist when he tries to integrate the evidence for megalithic astronomy into his picture of the society of the megalith-builders. At Stenness the astronomical theory implies a connexion between two sites separated chronologically by many centuries. Indeed much work on megalithic astronomy lumps together structures which the archaeologist sees no reason for connecting. One way to rectify the situation is chosen by Aubrey Burl (see his paper in this volume), who recommends studying the orientation of monuments which can be grouped together on archaeological grounds. This makes the task of the archaeologist easier, obviously, but unless archaeologists have said the last word on the connexions between monuments of different types and in different places, it is at least possible that useful evidence will be ignored by such a 'correct' approach. The extent to which archaeologically mixed or uncertain material can contribute is something considered by Dr Ruggles in his paper.

Another aspect of megalithic astronomy of importance in the wider context is its purpose. This can be classified in various ways, for example, as ritual, practical or scientific, though these terms have no significance which is universally agreed. But one concept which appears more and more to find favour is the idea that all the orientations for which the evidence is satisfactory can be understood as ritual orientations of no great precision. Again this seems a comfortable idea that makes as few demands on us as the construction of such orientations would have done on the megalith-builders. And yet it may not be quite so simple. In his paper in this volume Aubrey Burl draws attention to the orientations of the Clava passage-graves. Though the data is admittedly provisional, the apparent existence of orientations to the standstills of the moon implies that observations were made, recorded and recalled over a period of the order of 19

years. It seems unreasonable to suppose that this was done on the basis
of haphazard or sporadic observations; rather, there must have been a relatively systematic programme of observations, which no one observer may
have lived long enough to complete (Atkinson 1975). These orientations are
not particularly accurate, but the orientation of the avenue at Stonehenge
(see Atkinson's paper in this volume) indicates, in the simpler context of
a solstitial sightline, that considerably higher accuracy was attainable
and, perhaps, intended. It is a very small step from here to the solstitial
orientations of high accuracy, such as the north-west foresight viewed from
the central stone at Ballochroy. These have been a feature of Prof. Thom's
evidence ever since his earliest papers on megalithic astronomy, where, it
will be recalled, selection problems were relatively untroublesome. This
is not to say that such sightlines, if genuine, served a scientific purpose,
satisfying the curiosity of the megalithic observers about the motion of
the sun. But they are quite a long way from the picture of 'ritual orientations' which this phrase tends to evoke.

It might be hoped that the evidence of ethnoastronomy could
shed some light on the meaning of the astronomical orientations that we
find. Aubrey Burl's paper illustrates one of the few contacts which this
discipline has had with megalithic astronomy so far, and there seems little
prospect that this will improve. This state of affairs, and especially
what was seen as the reluctance of European archaeoastronomers to draw on
the evidence of ethnoastronomy, was the basis of some mild criticism from
the Americans at the Oxford conference. Perhaps one can understand why,
in the context of much American archaeoastronomy, the absence of any contact with ethnohistorical evidence should be viewed with suspicion, but the
immense value of Prof. Thom's work shows that, in the megalithic context,
the sanction of ethnoastronomy should not be seen as mandatory. In his
paper, Prof. Atkinson reminds us of the wise words of Jacquetta Hawkes,
who warned us that our view of Stonehenge is inevitably coloured by the
age we live in. If the Age of Science is followed by an Age of Ethnoastronomy, let us hope her words are not forgotten.

NOTES

1. A number of useful comments have been made about that paper.
Prof. Atkinson kindly pointed out that the balance of evidence clearly favours the supposition that hole F at Stonehenge is natural (p.S29), and
comments by Dr Ritchie and Prof. Thom have been incorporated within the
present text. Also, on p.S32, line 24, for 'Kintraw' read 'Kilmartin'.
2. The page number refers to the Fontana edition (London: 1970).

REFERENCES

Atkinson, R.J.C. (1975). Megalithic astronomy - a prehistorian's comments.
    J. Hist. Astron., 6, 42-52.
Barlai, K. (1980). On the orientation of graves in prehistoric cemeteries.
    Archaeoastronomy (Bulletin), 3, no.4, 29-32.
Brennan, M. (1980). The Boyne Valley Vision. Portlaoise: Dolmen Press.
Burl, H.A.W. (1976). The Stone Circles of the British Isles. New Haven and
    London: Yale University Press.
Burl, H.A.W. & Piper, E. (1979). Rings of Stone. London: Frances Lincoln.
Dobrzycki, J. (1963). Astronomiczna interpretacja prehistorycznych zabytków
    na terenie polski. Kwart. Hist. Nauk. Tech., 8, 23-7.
Freeman, P.R. (1975). Carnac probabilities corrected. J. Hist. Astron., 6,
    219.
Freeman, P.R. & Elmore, W. (1979). A test for the significance of astronom-
    ical alignments. Archaeoastronomy, 1, S86-S96.
Hadingham, E. (1981a). The lunar observatory hypothesis at Carnac: a recon-
    sideration. Antiquity, 55, 35-42.
    (1981b). (Book review). Archaeoastronomy (Bulletin), 4, no.2,
    44-5.
Hawkins, G.S. (1968). Astro-archaeology. In Vistas in Astronomy, ed. A.
    Beer, 10, pp. 45-88. Oxford: Pergamon.
    (1971). Photogrammetric survey of Stonehenge and Callanish. In
    National Geographic Society Research Reports (1965 Projects),
    ed. P.H. Oehser, pp. 101-8. Washington: Nat. Geogr. Soc.
    (1973). Beyond Stonehenge. London: Hutchinson.
Hawkins, G.S. & White, J.B. (1965). Stonehenge Decoded. New York: Double-
    day.
Heggie, D.C. (1981a). Megalithic Science. London: Thames & Hudson.
    (1981b). Highlights and problems of megalithic astronomy. Arch-
    aeoastronomy, 3, S17-S37.
MacKie, E.W. (1974). Archaeological tests on supposed prehistoric astronom-
    ical sites in Scotland. Phil. Trans. R. Soc. Lond. A, 276, 169-
    91.
    (1977). The Megalith Builders. Oxford: Phaidon.
Merritt, R.L. & Thom, A.S. (1980). Le Grand Menhir Brisé. Archaeol. J., 137,
    27-39.
Morrison, L.V. (1980). On the analysis of megalithic lunar sightlines in
    Scotland. Archaeoastronomy, 2, S65-S77.
Müller, R. (1936). Himmelskundliche Ortung auf Nordisch-Germanischem Boden.
    Leipzig: Curt Kabitzsch.
    (1970). Der Himmel über dem Menschen der Steinzeit. Berlin:
    Springer-Verlag.
Patrick, J.D. (1981). A reassessment of the solstitial observatories at Kin-
    traw and Ballochroy. In Ruggles & Whittle (1981), pp. 211-19.
Roy, J.-R. (1980). Comments on the astronomical alignments at Callanish,
    Lewis. J. Roy. Astron. Soc. Can., 74, 1-9.
Ruggles, C.L.N. (1981). A critical examination of the megalithic lunar obs-
    ervatories. In Ruggles & Whittle (1981), pp. 153-209.
    (1983). A reassessment of the high precision megalithic lunar
    sightlines. Part two: foresights and the problem of selection.
    Archaeoastronomy, 5, to appear.
Ruggles, C.L.N. & Whittle, A.W.R., eds. (1981). Astronomy and Society in
    Britain During the Period 4000-1500 B.C. Oxford: Brit. Archaeol.
    Rep.
Thom, A. (1954). The solar observatories of megalithic man. J. Brit. Astron.
    Ass., 64, 396-404.

Thom, A. (1955). A statistical examination of the meaglithic sites in Brit-
        ain. J. R. Stat. Soc. A, 118, 275-95.
        (1966). Megalithic astronomy: indications in standing stones.
        In Vistas in Astronomy, ed. A. Beer, 7, pp. 1-57. Oxford: Per-
        gamon.
        (1967). Megalithic Sites in Britain. Oxford: Clarendon Press.
        (1969). The lunar observatories of megalithic man. In Vistas
        in Astronomy, ed. A. Beer, 11, pp. 1-29. Oxford: Pergamon.
        (1971). Megalithic Lunar Observatories. Oxford: Clarendon
        Press.
Thom, A. & Thom, A.S. (1978a). Megalithic Remains in Britain and Brittany.
        Oxford: Clarendon Press.
        (1978b). A reconsideration of the lunar sites in Britain. J.
        Hist. Astron., 9, 170-9.
Thom, A.S. (1981). Megalithic lunar observatories: an assessment of 42 lunar
        alignments. In Ruggles & Whittle (1981), pp. 13-61.

ARCHAEOLOGY AND ASTRONOMY: AN ARCHAEOLOGICAL VIEW

J. N. Graham Ritchie

Royal Commission on the Ancient and Historical Monuments of Scotland

Abstract.   This paper examines some of the problems of the archaeological evidence from sites for which astronomical interpretations have been proposed;   these include problems of interpretation caused by the use of many prehistoric sites over a long period of time, and difficulties raised by the absence of a firm chronological frame-work. The absence of detailed gazetteers for many of the types of site under discussion is also stressed.   A number of excavations are used to contrast the material available to the prehistorian and to the astronomer.

Introduction

The astronomical interpretation of stone circles and standing stones has a long history; in the late seventeenth century, for example, Martin Martin, writing about the Ring of Brodgar and the Stones of Stenness in Orkney, records that 'Several of the Inhabitants have a Tradition that the Sun was worshipped in the larger, and the Moon in the lesser Circle' (1716, 365).   In recent years a scientific basis for interpretation of the archaeological evidence in astronomical terms has been proposed by Alexander Thom (Thom, A. 1967;   1971;   Thom & Thom 1978).   In 1965 he began his introduction to megalithic astronomy: 'Much has been written for and against the astronomical significance of the stone circles, stone alignments, etc., which are scattered throughout these islands and indeed much further afield.   There is, however, universal agreement that the erectors, herein called for convenience Megalithic man, marked the rising and setting points of the solsticial Sun' (Thom, A. 1965, 1).   In more recent books and papers he has also proposed, amid greater controversy, a series of lunar observatories. The present introduction was designed as a conference paper offering both a personal assessment of the range of sites for which astronomical interpretations have been proposed, and as a background to the more detailed papers that follow.   As delivered, it was intended to provide both a visual evocation of the range of material available and also

a body of evidence from which a number of general points about our understanding of the sites might be made.   In setting the archaeological
scene, a broadly chronological approach to the works of 'Megalithic
man' in Britain, highlighting some of the fundamental problems and the
inadequacies of the archaeological evidence, may be useful.

The structural use of large stones, particularly in northern
and western parts of the country, occurs at many stages in man's past;
thus we should no longer envisage any unity of megalithic 'culture'
or 'tradition' nor think in terms of 'Megalithic man'. Such a figure
was easier to imagine at a time when the neolithic and bronze ages in
Britain were thought to span a millennium and a half; but we now know
that the monuments span a period of about four millennia, and complex
relationships and the existence of independent traditions are altogether
more likely.   The use of timber for monumental structures in southern
and eastern parts of Britain is too readily overlooked if we are bemused
by the megalithic label.

Some of the earliest sites are the burial monuments known
as chambered tombs, built and used as collective burial-places for as
long as a millennium and a half, from 4000 BC to 2500 BC in broad terms
and in calibrated dates.   Earthwork circles, some containing rings of
standing stones or upright timbers, known as henge monuments, may belong,
again in very broad terms, to the millennium spanning 3500 BC to 2500
BC;   such henge monuments are presumably religious or ceremonial centres
for communities from the surrounding neighbourhood. Stone circles, stone
alignments, single standing stones and round cairns with monumental
kerbs are rather more difficult to date, but they form the bulk of the
monuments for which astronomical interpretations have been claimed.
Our vision of past societies depends, of course, not only on the evidence
of burial and ritual monuments, but also on that from settlement sites,
pottery and environmental data (see, for example, Megaw and Simpson
1979; Burgess 1980).

Chambered tombs

The structural evidence itself is not always easy to interpret
and excavation often reveals that more than one building period may
be involved;   many centuries after the initial construction of a site,
burials or ritual deposits may be inserted, attracted by the numinous
aura of impressive upright stones.   The multi-period nature of many
prehistoric monuments may be illustrated initially by two very different

chambered tombs: Wayland's Smithy in Oxfordshire and Achnacreebeag
in Argyll.   At Wayland's Smithy the primary structure was a small mound
with a complex mortuary enclosure containing partially articulated
burials (the skeletons in part at least in anatomical order) and
associated with two massive pits, which had formerly held substantial
split tree-trunks.   Later, but perhaps not much later, the small mound
was buried beneath a much larger one associated with a monumental
transepted megalithic tomb at a date of about 3500 BC.   At Wayland's
Smithy, the sequence is shown stratigraphically, one phase being sealed
by another (Atkinson 1965). At Achnacreebeag, two upstanding megalithic
structures were clearly visible before excavation and the question was:
were they contemporary and if not, which was the earlier?   The answer,
that the simpler structure was the earlier, and the small passage-grave
a later addition, was arrived at by the careful examination and planning
of the stones of the cairn, showing that the cairn enclosing the passage-
grave had been built at a later date on to one side of an existing round
cairn (Ritchie 1970).   This type of rather more oblique relationship
between structures is a constant feature of the excavation of stone
circles and is one that makes the working out of any sequence rather
more subjective, even on excavated sites.

         Can the Achnacreebeag sequence be extrapolated to the
unexcavated chambered tomb of Greadal Fhinn, situated near the tip of the
Ardnamurchan peninsula in Argyll, where very similar structural remains
survive?   Only excavation will tell; sadly, however, this site is still
included as a stone circle on many distribution maps, probably because
of its description as such on early Ordnance Survey maps, but there
is no doubt that it is a passage-grave within a round cairn (Henshall
1972, 358-60). This example may be used to stress the absence of reliable
descriptive gazetteers for many of the types of site under discussion,
except for chambered tombs themselves (in Scotland, for example, Henshall
1963 & 1972).   There are many lists of sites, but we lack gazetteers
for stone circles, alignments or cairns with detailed descriptions and
plans, and thus we lack a firm archaeological data-base in published
form, though there are of course good local or general studies (notably
Burl 1976).   All too frequently astronomers and archaeologists list
a site under different names and locate them in different ways (by
latitude and longitude or by National Grid Reference).   Only rarely
in the astronomical literature is there cross reference to archaeological

discussion - excavation reports, Royal Commission Inventories or Ordnance
Survey Record Cards for example;   thus there may be no account taken
of the past history of the site including the movement or re-arrange-
ment of stones.   This is important because stone structures of many
types in a collapsed or ruined state may superficially resemble 'stone
circles' or 'stone alignments'.   One illustration of this sort of problem
may be cited: The Eleven Shearers, Roxburghshire, described by Thom
as an alignment (1967, 149) is included in the Commission's Inventory
(pace MacKie 1975, 78-9) 'as nothing more than the grounders of an
ancient field-dyke' (RCAMS 1956, 194, no. 409), although such contrasting
interpretations are clearly a matter of opinion.   We lack complete
horizon-plots for most sites to compare with the selected fragments
that are thought to be astronomically significant.

Chambered tombs have indeed found their way into astronomical
discussions, Unival and Clettraval in North Uist, for example, although
the original nature of the tombs as burial places was not necessarily
recognised (for the archaeological interpretations see Moir 1980; Ruggles
& Norris 1980;   Atkinson 1981, 206; Ruggles 1981, 164).   But studied
as a group, chambered tombs do not appear to' have connections with
astronomical happenings, with a few important exceptions.   The monumental
cruciform passage-grave at Newgrange, in County Meath, is of interest
both because it is surrounded by a stone circle and because the
orientation of the passage permits the light of the midwinter rising
sun to illuminate the end recess of the chamber through an unusual
opening above the passage - the 'roof box' (Patrick 1974).   If the
surrounding stone circle is indeed an integral part of the original
plan, a date of about 3300 BC may be put forward for it on the evidence
of the radiocarbon dates, and it is thus among the earliest stone circles
known at present.   At Maes Howe, in Orkney, one of the most magnificent
of British passage-graves, a clear interest in celestial events may
also be seen in the orientation of the entrance passage, for the rays
of the setting sun illumine the rear wall of the central chamber at
the time of midwinter sunset.   The fact that the stone that blocked
the passage is not quite tall enough to fit to the top of the entrance
has also been brought into the discussion, and Welfare and Fairley suggest
that this was to allow a shaft of light to penetrate the chamber even
when the entrance was sealed (1980, 93).

The Stones of Stenness and the Ring of Brodgar

Recent excavations have shown that the use of the Maes Howe
class of tomb in Orkney is broadly contemporary with that of the stone
circle and henge monument of the Stones of Stenness - that is at a date
early in the third millennium BC. At Stenness, the absence of strati-
graphical relationships between the various structural elements is
apparent; pits A-E for example are probably as late as AD 500 or so
as a result of a radiocarbon determination from one of them (ad 519
± 150   SRR-352); (Ritchie 1976, 15). The circle is surrounded by the
ditch and bank of the henge, but it seems likely that they and the
rectangular stone setting at the centre of the site are broadly
contemporary - the evidence being provided more by symmetry than anything
else. Pottery from the bottom of the ditch and from within the central
setting is comparable to material from the chambered tomb of Quanterness,
11 km to the E, excavated by Colin Renfrew (1979). As there are also
similar radiocarbon dates from Quanterness we may be justified in making
the chronological equation between the use of Stenness and that of
Quanterness, and thus we know something of the burial places and of
the 'ceremonial' centres, be they for religious or secular activities
within the community. Similar radiocarbon dates and pottery within

Fig. 1. Stones of Stenness, Orkney: plan of structures
at the centre of the circle, square stone setting with the
position of a possible timber upright, sockets with packing
stones from which uprights have been removed, and remains
of possible timber setting (Ritchie 1976, fig. 4).

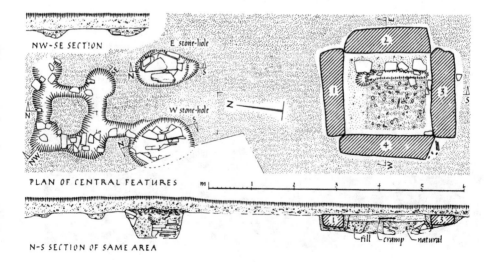

NW-SE SECTION   E stone-hole

W stone-hole

PLAN OF CENTRAL FEATURES   m

N-S SECTION OF SAME AREA   fill   cramp   natural

the same broad typological class known as 'grooved ware' found at the
settlement sites of Skara Brae and Rinyo shows that we can appreciate
the economic position of the users of such monuments more precisely
than perhaps anywhere else in Britain (see Clarke 1976 a and b).   An
example of the sort of re-organisation that may take place within stone
circles is illustrated by the setting to the north of the centre of
the site (Fig. 1), where two sockets dug to receive upright stones were
found;   the stones had been removed, but some of the original packing
was still in position.   The stones seem to have been associated with
a square timber setting, but sadly there is no helpful indication of
the date of the original construction or their deliberate removal.
It is surprising perhaps that no astronomical significance has ever
been postulated for what may have been an upright post within the central
stone setting at Stenness (Ritchie 1976, 12-13);   do we see here a
gnomon for the observation of shadows cast by the sun or is this a
 preliminary marker in the observation of the perturbation wobble of
the moon (Thom,A. S. 1981, 57)?

        The area round the Ring of Brodgar (Fig. 2), situated 1.5
km to the NW, has several suggested astronomical alignments, making

Fig. 2.   Ring of Brodgar, Orkney.

use particularly of the burial mounds around the henge and stone circle.
Recent excavation by Colin Renfrew has been confined to examination
of the ditch and possible bank, and the radiocarbon dates obtained do
little more than indicate that the ditch would still have been a conspicuous feature in the second half of the first millennium BC (Renfrew
1979, 39-43).   It is worth stressing that the Stenness dates should
not be used to indicate the period of use of the Ring of Brodgar, sister
monuments though they appear to be (Thom, A. S. 1981, 42-3), for the
dates obtained for various henge monuments in Britain cover a wide chronological span.   But it is not at all impossible on archaeological grounds
that the barrows round Brodgar are broadly of the date put forward by
Alexander and A. S. Thom for the use of the site as a lunar observatory
- that is between about 1600 BC and 1400 BC (Thom & Thom 1973;   1975;
1978, 122-37).   On the other hand it is wrong to imply that structures
can be 'dated by archaeoastronomy' (Baitey 1973, 399);   only a possible
period of use of a stone or circle can be so indicated.

This discussion of the Ring of Brodgar tellingly illustrates
an important difference between the sort of evidence archaeologists
and astronomers have at their disposal.   The relevant positions of all
lunar standstills from 2088 BC to 1307 BC as calculated from figures
provided by the Royal Greenwich Observatory for the various sightlines
round the Ring can be plotted (Thom & Thom 1975, 100, fig. 7).   This
reflects the rather more precise nature of the astronomical information
- available in chart form, susceptible of comparative precision even
when translated into past ages and various latitudes.   This is in stark
contrast to the silent stones of the archeological landscape, at periods
when there are no written records at all.   Examination of the literature
on such astronomical matters will show that many archaeologists use
the basis of their own excavations and survey to build up a body of
evidence, about which in their own minds they are reasonably certain,
and this policy has deliberately been continued, at least in part, in
this introduction.   Archaeological statements thus seem to be more
personal than astronomical ones and are based on individual interpretations of what is acceptable evidence;   it is very rarely possibly
to disprove the astronomical potential of any site on astronomical
grounds, and it is more often the validity of the archaeological evidence
that is in contention.   The more philosophical question of whether
ancient man is likely to have used standing stones as markers for

astronomical observation is not one that is possible to answer at all,
except perhaps from a statistical point of view, using both archaeological and astronomical observations that are acceptable to all.   It
is interesting that evidence for the use of features of the horizon
as foresights for the detailed observation of the moon does not seem
to be forthcoming from other societies in northern latitudes.

Standing Stones

Single standing stones occur widely throughout northern
and western Britain, sometimes isolated and remote, sometimes in apparent
association with other groups of site such as cairns or cists;   nor
are all standing stones of prehistoric date, as some have been set up

Fig. 3.   Try, Gulval, Cornwall: A, plan and section of
standing stone, cist and cairn. a, medium brown soft soil;
b, dark brown soft soil filling the socket from the menhir;
c, grey leached gritty soil;   d, chocolate-brown weathered
rab;   e, rab upcast of the pit dug to receive the cist;
 B, beaker from the cist (scale 1:3) (after Russell and
Pool 1964, 17, fig. 5 and 19, fig, 6, no. 7).

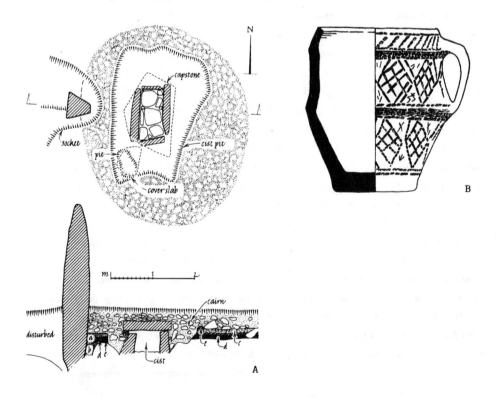

in comparatively recent times as route- or boundary-markers, scratching
posts for cattle, or indeed fence posts (as noted in Kansas by Myres
1965).   Only when there is evidence from within the backfilling of the
hole or socket that was dug into the ground to receive them, or strati-
graphical information about their sequence within a complex monument
is it possible to suggest a date for their construction.   Until recent
times it was only in remarkable cases that it was possible to be sure
of the date of the removal or destruction of a stone, as for example
the 'Barber's Stone' at Avebury, felled about AD 1320-5 accidentally
trapping an unfortunate barber-surgeon beneath it, or Odin's Stone in
Orkney broken up amid local consternation in December 1814.

         Excavations of standing stones include that at Try, Gulval,
in Cornwall (Fig. 3), for example, where beside a monolith there was
a cairn of stones, which covered a cist containing a burial associated
with a beaker vessel (Russell and Poole 1964).   The excavators describe,
however, that the upcast soil from the pit dug for the cist partly over-
lay the soil filling of the socket of the upright, and thus the stone
is the earlier feature;   the absence of any silting or weathering layer
suggested that there was only a short interval between the two operations.

         Fig. 4.   Maol Mor, Dervaig, Mull, Argyll: standing stones.

The handled beaker thus provides an approximate date for the erection of the menhir within the suggested chronology for such vessels at the time of any discussion, currently about the 18th to 17th century BC. An independent chronology provided by radiocarbon analysis, had suitable material existed, would of course be rather more satisfactory, and unlike typology, less susceptible to changes of archaeological fashion.

Many of the same archaeological problems occur over the dating of alignments of standing stones, though their potential for interpretation in astronomical terms is much greater; an example of a linear setting of stones from Mull illustrates this class of site (Fig. 4) - Maol Mor, Dervaig (Thom's Dervaig A). Excavation of the sockets of menhir and stone alignments have in fact done little more than stress a function related to burial or activities for which a dedicatory deposit of human cremated bone was appropriate. In primary association with the erection of one of the stones of the Ballymeanoch alignment, in Argyll, there were distinct deposits of bone against the NE and SE faces of the stone (5 gm in each), with a larger deposit of 90 gm on the SW side near the base of the socket (Barber 1978, 106-7). Within the socket of a pair of stones at Orwell, Kinross-shire, a double cremation deposit separated by a flat slab was probably part of the ceremonial at the time of the erection of the stone and the back-filling of the socket; the presence of burnt dog and pig bones with the lower cremation deposit may be thought to provide an impression of the original rituals of cremation. In the nineteenth century further cremation patches and cists were found round the stones (Ritchie 1974, 8-9). A date for some of these activities may be suggested by the excavation at Pitnacree, Perthshire, where at the top of the mound there was a standing stone, at the base of which there was a scattered cremation and a large amount of carbonised wood; this provided a radiocarbon date of 2270 bc ± 90 (GaK-602) (Coles & Simpson 1965, 38). The presence of burials associated with food vessels and cinerary urns indicates that the use of standing stones in this focal way may have continued into the mid-second millennium BC and beyond (Burgess 1980, 345-6), and this may have a bearing on their postulated use as astronomical markers. At Duntreath (or Blanefield),in Stirlingshire, the excavation was undertaken to examine the likelihood of the use of stones in an astronomical way (MacKie 1973; 1974, 187); Euan MacKie found a layer of ash and charcoal which suggested fires associated with the stones and which provided

a radiocarbon date of 2860 bc ± 270 (GX-2781), although MacKie stresses
that this was not conclusive (1977, 118).   MacKie's final caution may
be echoed;   Duntreath illustrates the dangers of assuming that all the
problems are solved simply because none are apparent in the small quantity
of hard evidence available   (1974, 187).   Excavations such as Orwell
and Duntreath are not perhaps exciting in themselves, but in the short
term they illustrate the sort of small-scale work that is needed to
build up a positive body of evidence about the dates and associations
of far more stones;   in the long term, however, only more extensive
work, as at Rhos-y-clegyrn, Pembrokeshire, will put such sites in a
proper context (Lewis 1974).

<u>Stone Circles</u> (Fig. 5)

The excavation of a stone circle may also raise problems
of stratigraphy and the relationship of the various pieces of evidence,
not least because it is difficult to know when a feature within a
a stone circle relates to an early phase of its construction or whether
it is a later addition - like the pits from the Stones of Stenness
mentioned earlier.   We may suppose, from the symmetry of the deposition,
that the cinerary urn from Sandy Road, Scone, in Perthshire, is broadly
contemporary with the setting up of the circle and that the radiocarbon

Fig. 5.   Lochbuie, Mull, Argyll: stone circle.

date of 1200 bc $\pm$ 150 (GaK-787) from charcoal within the urn refers
to the construction of the site itself, but this is little more than
an assumption, and the central position would also be a favoured one
for later depositions (Stewart 1966).

This demonstration of the problematic nature of the archaeological evidence does not disguise the fact that much of the material at
our disposal has a wide chronological span, certainly from 3000 BC to
at least the middle of the second millennium BC. Ruggles has illustrated
the variation in the setting lines for the Moon's upper and lower limbs over the
period 2500 BC to 1500 BC at Ballinaby, Islay (1981b, 753; compare Thom
& Thom 1978, 170, fig. 13.1). This sort of exercise undermines confidence
in the lunar lines proposed by the Thoms when one considers that on
archaeological grounds the sites must range (possibly randomly and
possibly with concentrations) from Stenness early in the third millennium
BC (on the evidence of the radiocarbon dates) perhaps to Fowlis Wester
in the middle of the second millennium BC (an archaeological estimation)
(Atkinson 1979, 101). The poster paper presented by the writer to
the conference took up the theme of site interpretation; this is of
paramount importance in providing a clear account of the archaeological
background. Stone circles, single standing stones and burial cairns
belong to distinctive categories of monument in some cases constructed
at different dates for different purposes, and in other cases forming
what appear to be small associated groups. Thus when used together
to provide a series of sites for which similar astronomical functions
are proposed, the archaeologist is bound to feel uneasy (Fleming 1975).
In an interesting exercise comparing the astronomical potential of
two structurally similar sites, Temple Wood and Barbreck in Argyll, Jon
Patrick found that no similarities in astronomical use could be postulated
for the latter (1979), and this may be thought to cast doubt about the
lines observed at the former (Heggie 1981b, 186).

Stone circles themselves can be broken down into several
classes both on typological and geographical grounds; including the
'entrance-circles' of Cumbria, the 'recumbent stone-circles' of SW
Ireland, the 'four-posters' of Perthshire and NE Scotland (with some
outliers), but perhaps the best known being the 'recumbent stone-circles'
of NE Scotland, recently studied in detail by Aubrey Burl, including
an examination of their archaeoastronomical potential (1980). In the
past our understanding of the various groups has been bedevilled by
the fact that many stone circles proved a magnet for nineteenth century

antiquaries with indiscriminate excavation methods, and though we may
know what was discovered in them, the various relationships are less
than certain.   In the north, the excavations of Howard Kilbride-Jones
at Loanhead of Daviot and Cullerlie helped to change the format of
reporting, but sadly Loanhead was already severely disturbed (1935).
After the War the excavations of Stuart Piggott at Cairnpapple,
the Clava sites, and latterly at Croft Moraig significantly increased
our knowledge;   at the latter site the interesting sequence from
timber monument to stone circle was demonstrated (Piggott and Simpson
1971).   Stuart Piggott has recorded the visit of Ludovic Mann, the veteran
Scottish antiquary, to Cairnpapple in 1948;   Mann's only words on seeing
the plan of the stone holes of the henge were 'It's an egg, Professor,
and that's the beginning of a new era!'   Today such a view would no
longer be thought altogether eccentric, as Alexander Thom has convincingly
shown that stones were deliberately set out in several geometric shapes:
the flattened circle, the ellipse, the egg, and the complex circle
probably using a standard unit of length the 'Megalithic Yard' (Thom
& Thom 1978, 18, fig. 3.1).

<u>Cairns</u> (Figs. 6 & 7)

One distinct class of cairn is of particular interest to
those working on astronomical interpretations because examples include
cairn B at Kintraw, the cairns at Monzie, Ballymeanoch and one of the
pair of sites at Fowlis Wester.   The excavation of the small cairn at
Strontoiller in Argyll (Thom's site Loch Nell Al/2) showed that large
erratic boulders formed the contiguous kerb-stones of a low cairn, with
a fragmentary cremation burial in the lowest levels;   scatters of white
quartz pebbles were found round the kerb stones (Ritchie 1971).   Kerb
cairns are clearly closely related to stone circles at Lochbuie,
Strontoiller and Temple Wood;   to linear settings of standing stones
at Ballymeanoch and at Ardnacross, on Mull;   and they are often associated with   single outlying standing stones, for example at
Strontoiller and Achacha.   Astronomical inferences have been drawn for
some, though not all, the examples of this class, frequently in the
past described as stone circles, though this can no longer be an accurate
label;   echoing Aubrey Burl's recent plea for the study of groups of
monuments, it would be instructive to see if from an astronomical point
of view a consistent use is forthcoming for this class as a whole.   Kerb
cairns as a group may be related to the recumbent stone circles of NE

Fig. 6. Ardnacross, Mull: standing stone and cairns.

Fig. 7. Strontoiller, Argyll: kerb cairn.

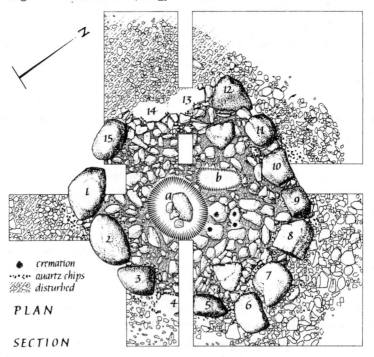

cremation
quartz chips
disturbed

PLAN

SECTION

Scotland and to the ring-cairn tradition, but the nature of such cross-country contacts is impossible to determine archaeologically, though very similar cairns are found in the north-east - perhaps most remarkably at Cullerlie. Some indication of the date of the kerb-cairn class is provided by the excavations at Claggan, Morvern, Argyll, where there are two radiocarbon determinations from charcoal associated with cremations: 975 bc ± 50 (SRR-284) and 1058 bc ± 40 (SRR-285) indicating a date around 1200 to 1300 BC in corrected terms (Ritchie et al. 1975).

Conclusions

The problems of archaeological interpretation have been deliberately stressed for several reasons - one is that the archaeological evidence at present formulated, at least in terms of published material, cannot provide the firm basis for the sort of examination that is so clearly needed for an independent evaluation of the astronomical hypotheses. Secondly, the chronological span of stone circles and cairns (which are more susceptible to modern dating techniques than single standing stones or alignments) means that a greater appreciation of the various typological and chronological groups is essential if the astronomical possibilities are to be slotted into a broad pattern. The position and orientation of many of the sites had almost certainly been established by 1700 BC to 1500 BC, by the time that the more precise astronomical observations are postulated.

Douglas Heggie posed an important question in his introductory paper to the conference: must unexcavated sites be excluded from astronomical consideration until they have been excavated? (1981a, 29). The archaeological purist might say 'yes', and enough has been said to underline the uncertainties of the archaeological evidence, but this is not altogether a practical answer in view of the small number of sites that will in fact be excavated even by the end of the century; there may, however, be grades of uncertainty. The superficially most complex sites (chambered tombs and stone circles) were probably in use, and being constructed over a long period of time, and in an unexcavated state may show merely the surviving features of the final stage of building or use. Rather simpler sites (linear setting of stones and cairns) are not as liable to change shape radically on excavation, and may be more helpful even in a raw state, though their chronological position will remain uncertain. It is from an interpretation of the field evidence that is accepted by all within an agreed chronological frame-

work (despite all the problems that this raises for the archaeologist)
that the subject must proceed (Ritchie 1980, 98) rather than on its
present parallel paths.   In his discussion of future development Heggie
stresses the importance of a rigorous attention to a statistical approach
(1981a, 34; 1981b) and it is likely that only with joint programmes will
the subject be taken forward: archaeological responsibility for site-
interpretation, classification and possible chronological niche; site-
survey and horizon-plotting to the highest standards within an archaeological and astronomical framework;   statistical evaluation of the
groups of sites with potential astronomical indicators.   Such a process
would also lead to the publication of detailed site-gazetteers the
absence of which so hampers further research.

Alexander Thom's work has made many prehistorians re-assess
their views on the likely achievements of early societies in theoretical
fields; the architectural or structural achievements can be appreciated
in the field (stone circles, or chambered tombs), the high standard
of craftmanship be it in pottery or metalwork is obvious, but the role
of 'science' in society is difficult to assess.   Thom's work has,
however, had an important effect on the ways in which we evaluate our
material, whether in wide-ranging theoretical constructs or in a more
traditional approach to the society of prehistoric man (Baitey 1973,
415-16;   MacKie 1977;   Ellegård 1981;   Heggie 1981b).   The careful
surveys by Alexander Thom have also been a spur to more meticulous
work in archaeological recording, the fruits of which are already
appearing (Burl 1980;   Ruggles 1981a), but the rough north points of
many published archaeological plans mean that the sites will have to
be re-surveyed before their astronomical potential can be properly
assessed (see also Baitey 1973, 400).   The method of presentation both
of negative evidence and of horizon plots that have proved to be without
astronomical significance may also have to be reconsidered; the reaction
of Owen Gingerich, an astronomer, to Thom's hypotheses, following his
own field-examination of the sites, as 'pyschologically devastating'
to a tentative acceptance of precision astronomy in ancient Britain
(in Ellegård 1981, 117) is particularly telling.   Thus if to the present
writer a not proven verdict remains over the question of detailed astronomical observation in prehistoric times, this does nothing to diminish
Alexander Thom's contribution to our approach to past societies.

Acknowledgements and copyright of illustrations

The writer is indebted to the following for assistance during the preparation of this paper: Anna Ritchie, John Barber, D. C. Heggie, Linda Louden, Stuart Piggott, A. MacLaren, I. G. Scott, J. G. Scott, J. N. Stevenson, and the photographic staff of the Commission.

The figures are Crown Copyright, Royal Commission on Ancient Monuments, Scotland with the following exceptions: Fig. 2. Crown Copyright, Scottish Development Department (Ancient Monuments), under whose auspices the excavations at the Stones of Stenness (Fig. 1) were undertaken; Fig. 3A, re-drawn with kind permission Miss V. Russell and P. A. S. Pool ; Fig 3B after Russell and Pool 1964, fig. 6 , no. 7.

References

Atkinson, R. J. C. (1965). Wayland's Smithy, Antiquity, 39, 126–33.

Atkinson, R. J. C. (1979). The Thoms' new book (review of Thom and Thom 1978). Archaeoastronomy, 1, 99–102.

Atkinson, R. J. C. (1981). Comments on the archaeological status of some of the sites. In Ruggles and Whittle 1981, 206–9.

Baitey, E. C. (1973). Archaeoastronomy and ethnoastronomy so far. Current Anthropology, 14, 389–449.

Barber, J. W. (1978). The excavation of the holed-stone at Ballymeanoch, Kilmartin, Argyll, Proc. Soc. Antiq. Scot., 109, 104–11.

Burgess, C. (1980). The Age of Stonehenge. London: Dent.

Burl, A. (1976). The Stone Circles of the British Isles. New Haven and London: Yale University Press.

Burl, A. (1980). Science or symbolism: problems of archaeoastronomy. Antiquity, 54, 191–200.

Coles, J. M. & Simpson D. D. A. (1965). The excavation of a neolithic round barrow at Pitnacree, Perthshire, Scotland. Proc. Prehist. Soc., 31, 34–57.

Clarke, D. V. (1976a). The neolithic village at Skara Brae, Orkney: 1972–73 excavations: an interim report. Edinburgh: H.M.S.O.

Clarke, D. V. (1976b). Excavations at Skara Brae: a summary account. In Settlement and Economy in the Third and Second Millennia B.C. eds. C. Burgess and R. Miket pp. 233–50. Oxford: British Archaeological Reports, British Series no. 33.

Ellegård, A. (1981). Stone age science in Britain? Current Anthropology, 22, no. 2, 99–125.

Fleming, A. (1975). Megalithic astronomy: a prehistorian's view.

Nature, 255, 575.

Heggie, D. C. (1981a). Highlights and problems of megalithic astronomy.
    Archaeoastronomy, 3, 17-37.

Heggie, D. C. (1981b). Megalithic Science. London: Thames and Hudson.

Henshall,A. S. (1963 & 1972). The Chambered Tombs of Scotland. Edinburgh:
    Edinburgh University Press.

Kilbride-Jones, H. E. (1935). An account of the excavation of the stone
    circle at Loanhead of Daviot and of the standing stones
    of Cullerlie, Echt, both in Aberdeenshire, on behalf of
    HM Office of Works. Proc. Soc. Antiq. Scot., 69, 168-223.

Lewis, J. M. (1974). Excavations at Rhos-y-clegryn prehistoric site,
    St Nicholas, Pembs. Archaeol. Cambrensis, 123, 13-42.

MacKie, E. W. (1973). The standing stones at Duntreath. Current Arch-
    aeology, 2, no. 5, 279-83.

MacKie, E. W. (1974). Archaeological tests on supposed prehistoric
    astronomical sites in Scotland. Phil. Trans. R. Soc. Lond.,
    A. 276, 169-194.

MacKie, E. W. (1975). Scotland: An Archaeological Guide. London:
    Faber and Faber.

MacKie, E. W. (1977). Science and Society in Prehistoric Britain.
    London: Elek.

Martin, M. (1716). A Description of the Western Islands of Scotland.
    London: Bell. Reprinted 1970, Edinburgh: Mercat Press.

Megaw, J. V. S. & Simpson, D. D. A. (1979). Introduction to British
    Prehistory. Leicester: Leicester University Press.

Moir, G. (1980). Megalithic science and some Scottish site plans:
    Part I. Antiquity, 54, 37-40.

Myres, M. T. (1965). Standing field stones in Kansas. Antiquity, 39,
    136-7.

Patrick, J. (1974). Midwinter surise at Newgrange. Nature, 249, 517-
    19.

Patrick, J. (1979). A re-assessment of the lunar observatory hypothesis
    for the Kilmartin Stones. Archaeoastronomy, 1, 78-85.

Piggott, S. & Simpson D. D. A. (1971). Excavations of a stone circle
    at Croft Moraig, Perthshire, Scotland. Proc. Prehist. Soc.,
    37, 1-15.

R.C.A.M.S. (1956). Royal Commission on the Ancient and Historical
    Monuments of Scotland. An Inventory of the Ancient

and Historical Monuments of Roxburghshire. Edinburgh: H.M.S.O.

R.C.A.M.S. (1980). Royal Commission on the Ancient and Historical Monuments of Scotland. Argyll: an Inventory of the Ancient Monuments, 3, Mull, Tiree, Coll and Northern Argyll. Edinburgh: H.M.S.O.

Renfrew, C. (1979). Investigations on Orkney. London: Society of Antiquaries of London, Report of the Research Committee, no. 38·

Ritchie, J. N. G. (1970). Excavation of the Chambered Cairn at Achnacreebeag. Proc. Soc. Antiq. Scot. 102, 31-55.

Ritchie, J. N. G. (1971). Excavation of a cairn at Strontoiller, Lorn, Argyll. Glasgow Archaeol. J., 2, 1-7.

Ritchie, J. N. G. (1974). Excavation of the stone circle and cairn at Balbirnie, Fife. Archaeol. J., 131, 1-32.

Ritchie, J. N. G. (1976). The Stones of Stenness, Orkney. Proc. Soc. Antiq. Scot., 107, 1-60.

Ritchie, J. N. G. (1980). Review of Thom and Thom 1978. J. British Archaeol. Ass., 133, 97-8.

Ritchie, J. N. G. et al. (1975). Small cairns in Argyll: some recent work. Proc. Soc. Antiq. Scot., 106, 14-38.

Ruggles, C. (1981a). A critical examination of the megalithic lunar observatories. In Ruggles and Whittle 1981, 153-209.

Ruggles, C. (1981b) Prehistoric astronomy: how far did it go? New Scientist, 90, 750-3.

Ruggles, C. L. N. & Norris, R. P. (1980). Megalithic science and some Scottish site plans: Part II. Antiquity, 54, 40-3.

Ruggles, C. L. N. & Whittle, A. W. R. eds. (1981). Astronomy and Society in Britain during the Period 4000-1500 B.C. Oxford: British Archaeological Reports, British Series no. 88.

Russell, V. & Pool P. A. S., (1964). Excavation of a menhir at Try, Gulval. Cornish Archaeol., 3, 1964, 15-26.

Stewart, M. E. C. (1966). Excavation of a circle of standing stones at Sandy Road, Scone, Perthshire. Trans. & Proc. Perthshire Soc. Nat. Sci., 11, 7-23.

Thom, A. (1965). Megalithic astronomy: indications in standing stones. In Vistas in Astronomy, 7, ed. A. Beer, pp. 1-57. Oxford: Pergamon Press.

Thom, A. (1967). Megalithic Sites in Britain. Oxford: Oxford University Press.

Thom, A. (1971). Megalithic Lunar Observatories. Oxford: Oxford University Press.

Thom, A. & Thom, A. S. (1973). A megalithic lunar observatory in Orkney: the Ring of Brogar and its cairns. J. Hist. Astronomy, 4, 111–23.

Thom, A. & Thom, A. S. (1975). Further work on the Brogar lunar observatory. J. Hist. Astronomy, 6, 100–14.

Thom, A. & Thom, A. S. (1978). Megalithic Remains in Britain and Brittany Oxford: Clarendon Press.

Thom, A. S. (1981). Megalithic lunar observatories: an assessment of 42 lunar alignments. In Ruggles and Whittle 1981, 13–61.

Welfare, S. & Fairley, J. (1980). Arthur C. Clarke's Mysterious World. London: Collins.

THE STATISTICAL APPROACH

P.R. Freeman
Department of Mathematics,
University of Leicester, LE1 7RH, England

Abstract. A brief summary is given of the need for a
statistical approach to assessing most archaeoastronomical
data and theories. The various ways in which selection biasses
can lead to misleading conclusions, or at least seriously
diminish the value of observed data, are described and a few
possible techniques for making allowances for them are
suggested. A set of three idealistic rules is stated in the
hope of improving future observational work. Finally a brief
analysis is made of a particular set of data that satisfies
nearly all the requirements for lack of bias.

The subject of statistics has, on the whole, a pretty bad
public image. Even among scientists it is usually regarded as a necessary
evil, a hoop through which one's data must jump before getting accepted
for publication. Archaeoastronomers don't even seem to have got this far,
since most papers and talks at conferences seem oblivious of the need for
taking any kind of statistical view of the data presented. I very much
hope that two recent excellent works (Heggie 1981 a,b) will rapidly change
this attitude and I strongly recommend all readers of this article to
consult both these references. They make nearly all the points that I
should have wished to make, so I need not repeat them here.

We do not need statistics to help assess the evidence for
midsummer sunrise at Stonehenge or midwinter sunrise at Newgrange. Most
people will agree that the contexts are such as to virtually eliminate the
possibility that what we see today is a coincidence, unintended by the
original builders. But most sites are not like this, and when considering
a line of two or three standing stones, or even a single stone that does
not clearly indicate any direction in particular, we must always put any
claim for astronomical observation there to the test of seeing how much of
a coincidence is involved. There are several ways of thinking about this,
but perhaps the simplest is to say "Suppose I stood here with a pointer
smoothly pivoted about its centre. If I spun it, what are the chances that

it would come to rest pointing in a direction to which I could give some
sort of astronomical interpretation?" To answer this, I have to compile
a list of all astronomical events that I should find of interest and this
can be long, including solar solstices and equinoxes, lunar major and
minor standstills, planetary and stellar risings and settings, and so on.
The chances of my pointer indicating one of these, to within some limit of
error which again I must specify, at any one site, can be quite high. If
so, the evidence for astronomical intention at that site alone is weak,
but much stronger evidence can be obtained by considering as large a
number of sites as possible. If the number showing possible astronomical
intention is found to be much greater than the number that would be
expected by chance coincidence alone , then a statistically convincing case
will have been made. If not, or if no attempt is made even to assess the
evidence in this way, then no self-respecting scientist will take the
results seriously and the subject of archaeoastronomy will be left even
more than it is now to the gullible and the lunatic fringe.

As a statistician I have to work with whatever data other
people choose to collect. But as Heggie points out, the valid statistical
analysis is by far the easiest stage and very few sets of data get through
all the previous stages which are needed to qualify them as validly
analysable at all. These stages are nearly all to do with subjective biasses
and some of them are listed below. The very nature of the subject makes
several of them extremely difficult to avoid, or to deal with in ways which
most people will regard as fair, but it has to be said that many workers
have made far too little attempt to avoid them at all and the credibility
of their results has decreased accordingly. I shall now state aphoristically
a few very general instructions on how to deal with subjective bias.

RULE 1 OBSERVE EVERYTHING

The choice of what to observe is, I suspect, the biggest single
source of bias. We all know that asking a sample of unemployed manual
workers for their views on the performance of the current government is
likely to give a very misleading idea of the general opinion across the
whole country, yet similar elementary mistakes abound throughout the
archaeoastronomy literature. It is, I suppose, natural for those people who
firmly believe in the astronomical interest of ancient cultures to go and
survey those sites which look most likely to support their belief. When
at a site, it is equally natural for them to only survey those aspects of

it which fit in with their beliefs, ignoring all other data which they
regard as irrelevant but which a dispassionate reader must have in order
to assess their work. For example Thom and Thom (1981) surveyed a site in
Sutherland, consisting of a single menhir about 5 feet wide and 1 foot
thick. They surveyed the azimuth of this stone, in so far as it can be
said to have one, finding it to be 207°. They then surveyed very
accurately the horizon profile for azimuths between 195½° and 199½° only,
and after their usual precise calculations conclude that the site was
used for observing moonset in a small "notch" at major standstill. This
work is, of course, useless as evidence until someone else visits the
site and surveys the whole horizon profile so that the "random pointer"
concept can be applied. Notice here also how the actual megalithic hardware
plays virtually no part in the archaeoastronomical superstructure. In
reading the paper one must ask oneself about all the other possible places,
unmarked by any megalithic remains, where one might have stood and
observed some equally interesting rising or setting. I should like to
applaud here the work of the Pontings in recording so carefully complete
horizon profiles at Callanish, and I hope to do some statistical work with
these in the near future.

In this general area it is worth commenting on the various
attempts there have been to list criteria for what megalithic remains
should be agreed to be potential direction indicators. Thom (1955)
defined a set in his earlier work, but his later (1967) set depends on
subjective judgements. Hawkins (1968) said that only man-made markers
should qualify, that alignments should be postulated only for a homogeneous group of markers, and that all possible alignments at a site must
be considered. Cooke et al.(1977) give a hierarchal list of five classes
of alignments, ranging from those along 3 or more stones, through those
along only 2 stones down to indications by flat faces of a single menhir.
The Thoms' most recent work (1981) has clearly moved far from all these
criteria since they state "... the function of the backsight is to mark
a position, not a direction." My own attitude is that worthy as these
attempts at systematisation are, it is probably impossible to achieve a
set of criteria that can be generally agreed and it is not crucial anyway. The really important requirement is that each alignment should be
surveyed so as to record not only its azimuth, but the accuracy to which
that azimuth can be indicated by the components of the alignments themselves. It is extremely relevant that a pair of standing stones can only

indicate a direction to within about $\pm \frac{1}{2}°$ even when they are slender and
spaced well apart, while, for example, the axis of symmetry of a recumbent
stone circle or a flat-faced single menhir can get within only $\pm 2°$ or $3°$,
and a shapeless single menhir achieves the ultimate $\pm 180°$.

RULE 2   REPORT ALL YOU OBSERVE

Some selective judgement certainly has to be made when
preparing field work for publication, but any that chooses those observational aspects which support a particular theory in preference to those
that do not is obviously misleading. We often see survey plans on which
astronomical lines have been superimposed, for example Hawkins (1965) on
Stonehenge and Thom & Thom (1973) on Brogar. Much time has to be spent
thinking about all the other lines which might have been chosen but
weren't because they didn't indicate anything interesting.

Equally, all workers should report the accuracy to which their
measurements were made, and the discrepancies between indicated alignments
and theoretical ones. As a statistician I believe in error (indeed, I make
my living from it) and am distinctly unimpressed by phrases like "exactly
in line with", "spot on". Often a bald statement implying exact indications
turns out to cover up discrepancies of several degrees.

RULE 3   STATE CLEARLY THE RELATION BETWEEN HYPOTHESES AND DATA

The whole Popperian philosophy of science, in which a
hypothesis is constructed to be consistent with all known data, an
experiment is designed to test the hypothesis on new data, is then
performed and the data are assessed for consistency or conflict with the
hypothesis, is almost totally ignored in archaeoastronomy. This is not
only because it is an observational rather than experimental science (so
is ordinary astronomy, after all), but because it is much more difficult
to adopt this philosophy than to ignore it, and because many workers
(including the Thoms) seem to be unable to see any virtue in it at all.

The dangers of choosing a hypothesis to test after having
inspected a set of data and found something unusual in it are well known.
A classical example is the 18th century debate about whether the fixed
stars are randomly arranged over the celestial sphere. It was agreed that
the existence of the Pleiades was convincing proof of non-randomness,
since the probability of finding 6 stars all of magnitude 6 or greater all
within 49' of arc of each other under a purely random arrangement is only 1

in 33,000. The obvious counter-argument is that many purely random arrange-
ments will display some striking feature (3 stars nearly in a row, 4 stars
nearly forming a parallelogram, a group reminding you of a pussy cat, etc.)
which if then selected and tested for would also have a very small chance
of occurrence.

It is, I think, possible to push this argument too far, since
many important advances in science have come from such fortuitous noticing
of striking patterns in data, but all due allowance for them must be made
if at all possible. The Thoms have, after all, always reported their
surprise on constructing histograms of circle diameters or alignment
declinations without ever commenting on what other kinds of histograms
they would have found just as surprising.

Even more dangerous, of course, is adapting hypotheses from one
dataset to another. This new site shows no sign of solar or lunar indication? Never mind, try the planets, some stars, a few different dates,
something is bound to work. One can only guess how much of this kind of
thinking goes on behind the scenes as it is so easy to cover up in papers.
My personal precaution is to hold a hierarchy of plausible targets at the
back of my mind, ranging from solar solstices through solar equinoxes, lunar
major and minor standstills and the lunar wobble down to stars of decreasing visual magnitudes. The statistical significance of any set of
claimed alignments is then assessed using all targets in this list at or
above the lowest level of those that are claimed.

If allowing data to suggest hypotheses is potentially misleading so also is the converse sin of allowing hypotheses to dictate
what data are collected. This leads to beautifully circular arguments,
as in the Thoms' work on Avebury (1976) where the existence of a megalithic yard of 2.720 feet and knowledge of Pythagorean triangles is
assumed in "reconstructing" the geometry, and a least squares fit is then
used to reconfirm a megalithic yard of 2.722 feet. If, on the other hand,
a least-squares fit is done without preconceptions (Freeman, 1977) no
evidence of either a megalithic unit nor of a Pythagorean triangle is
found. The archaeoastronomical equivalent of this is found in the Thoms'
work (1971) on Er Grah, where lunar observations are assumed, the
directions in which to look for possible backsights are calculated and
required stones then located, several of mightily unimpressive size.
There are, of course, so many standing stones in the region that the
lunar hypothesis when forced upon the site is almost bound to lead to

self-fulfilling prophesy (see Atkinson 1975, Freeman 1975).

An example

Lest it be thought that all these objections to widespread
current practice set standards which it is impossible for any dataset
to satisfy, I conclude with one which, though not perfect, is pretty
good. This is a set of surveys of 31 recumbent stone circles in Cork
and Kerry conducted by Dr Jon Patrick. The sites are archaeologically
and culturally homogeneous (I should perhaps emphasise the importance
of this, even though it is rather outside the scope of this paper).
Although there are probably as many sites again still waiting to be
surveyed, there is no reason to believe that the selection of sites
has led to astronomical bias. For each site, Patrick has recorded

(a) the azimuth of the axis of symmetry, assuming a consistent use
of the direction from portals to recumbent,
(b) the inherent accuracy to which that azimuth can be determined
(c) the latitude of the site,
(d) the horizon altitude in the direction of the indicated line,
These form the input data to a computer program by Freeman and Elmore
(1978), the output from which forms fig.1. The horizontal axis is the
allowable discrepancy a between indicated declination and the "target"
ones - here taken to be solar solstices and equinox, rising and setting,
upper and lower limbs - while the vertical axis is the value of the test
statistic z. Since on the "random pointer" hypothesis this should have a
standardised Gaussian distribution, with negative values denoting more
coincidences than by chance, we need values of z below -2, and
preferably below -3, to provide any evidence of astronomy worth
considering. Here we can see there is no evidence for sun observations
at all. The full results from this dataset will, we hope, be published
elsewhere.

Finally, I am always most interested in hearing from anyone
who has similar homogeneous, unbiassed data suitable for statistical
analysis.

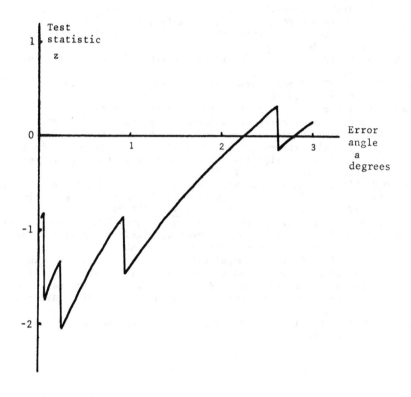

Fig.1.   Statistical analysis of Irish recumbent

stone circles.

References

Atkinson, R.J.C. (1975). Megalithic astronomy - a prehistorian's comments.
        J.Hist. Astron.,6, 42-52.
Cooke, J.A., Few, R.W., Morgan, J.G., & Ruggles, C.L.N. (1977).
        Indicated declinations at the Callanish megalithic sites.
        J.Hist. Astron.,8, 113-33.
Freeman, P.R. (1975). Carnac probabilities corrected. J.Hist. Astron.,6,
        219.
Freeman, P.R. (1977). Thom's survey of the Avebury ring. J.Hist.
        Astron.,8, 134-6.
Freeman, P.R. & Elmore, W. (1979). A test for the significance of
        astronomical alignments. Archaeoastronomy, 1, S86-S96.
Hawkins, G.S. (1965). Stonehenge decoded. London, U.K: Souvenir Press.
Hawkins, G.S. (1968). Astro-archaeology. Vistas in astronomy,10, 45-88.
Heggie, D.C. (1981a). Highlights and problems of megalithic astronomy.
        Archaeoastronomy 3, S17-S37.
Heggie, D.C. (1981b). Megalithic Science, London, U.K: Thames and Hudson.
Thom, A. (1955). A statistical examination of the megalithic sites in
        Britain. J.Roy. Statist.Soc.A, 118, 275-295.
Thom, A. (1967). Megalithic sites in Britain. Oxford, U.K: Clarendon Press.
Thom, A. & Thom, A.S. (1971), The astronomical significance of the large
        Carnac menhirs. J.Hist. Astron.,2, 147-60.
Thom, A. & Thom, A.S. (1973). A megalithic lunar observatory in Orkney:
        the ring of Brogar and its cairns. J.Hist. Astron.,4, 111-23.
Thom, A., Thom, A.S. & Foord, T.R. (1976). Avebury (1): a new assessment
        of the geometry and metrology of the ring. J.Hist. Astron.,7,
        183-92.
Thom, A. & Thom, A.S. (1981). A lunar site in Sutherland. Archaeoastronomy
        3, S71-S73.

STATISTICAL AND PHILOSOPHICAL ARGUMENTS FOR THE ASTRONOMICAL
SIGNIFICANCE OF STANDING STONES WITH A SECTION ON THE SOLAR
CALENDAR

A. Thom
Thalassa, The Hill, Dunlop, Kilmarnock, KA3 4DH.

A.S. Thom,
The Hill, Dunlop, Kilmarnock, KA3 4DH.

Abstract. Stressing the importance of the lunar bands, the
authors show that within each of these bands, the histograms
of declination give strong support to the lunar hypothesis,
because of the clustering of the alignments round the expected
values. Recently, while considering which lunar band edges
would be favourable or unfavourable for the erectors, the
authors have discovered a way of presenting the data published
up to 1978. This Favourable/Unfavourable objective
consideration gives strong support to the hypothesis, since
nothing was known of it over the decades when the surveys
were made. The method shows a low probability (1 in 433)
that the lunar alignments are occurring by chance alone. By
reasoning from the viewpoint of the erectors, the authors
give several supporting philosophical arguments which by their
very nature can not be used statistically; e.g. the warning
positions which could well have been occupied by an assistant
observer to tell the row of observers at the backsight of the
imminent Moonrise. A revised histogram of solar lines is
presented with the 16-month calendar based on a 365/366 day
year. It is suggested that the erectors had two reasons for
recording the Moon's movement; (a) eclipse prediction
assisted by their solar calendar and (b) scientific curiosity.

1 INTRODUCTION

Our knowledge of megaliths can come only from the remains themselves. These consist of standing stones, their positions relative to one
another, to tumuli, to mounds, and to nearby tracks on the ground. In
addition some of the stones have cup and ring markings. It is generally
accepted that some of the standing stones are arranged to indicate the
rising/setting points of the solstitial Sun, but why stop with the
solstices, and why not consider the Moon as well? The types of foresight
used at various places are detailed in Thom & Thom (1980 b).

Accurate surveying for declination. To take the subject further we must
make accurate surveys of the stones. Standing at the backsight, i.e. at
or near a standing stone, we measure by theodolite the altitude and azimuth

of the foresight. For the azimuth, observation of the Sun is necessary,
using an accurate watch. Knowing the latitude of the backsight we can then
determine the required azimuth by spherical trigonometry. From the measured
altitude and azimuth, again using spherical trigonometry, and our knowledge
of refraction, we calculate the declination of the foresight and so we know,
to within a minute of arc, the declination of any heavenly body that rose
or set on the foresight. We then examine the declinations thus obtained.
If the standing stones have nothing whatever to do with astronomy then
the declination will be spread at random throughout the range. If on the
other hand the declinations fall into clumps around significant points then
we know that we are dealing with stones erected for astronomical use. Up
to 1967 we had observed about 260 declinations. These are shown plotted in
Fig. 1 which is taken from Thom (1967). It is evident that the points
group around germane declinations. Our critics say that we perhaps
unintentionally use subjective methods in selecting our foresights but we
can not have done this every time. We consider that our choice was not
subjective, and this is where the matter rests at present. We have several
times stated that megalithic man was able to detect the small nine-minute
perturbation Δ of the lunar orbit. This needs much more careful consideration and in our analysis we must use much more sophisticated methods.

2 STATISTICAL VIEWPOINT

        In our long investigation into the astronomy of Megalithic Man
the greatest difficulty all along has been to decide just when a line should
be neglected. On the one side people are asking for evidence, on the other,
statisticians are waiting to point out any irregularities. We try to use
objective criteria. No matter what terms of reference are decided at the
beginning, however, there are always borderline cases. We want to produce
an answer to the question: Did Megalithic Man really use advanced astronomical methods for calendrical purposes, for eclipse prediction, or had
he perhaps purely scientific reasons? We wish to find the answer in a form
that will be acceptable to archaeologists but we find much greater
difficulty in satisfying statisticians who wish to apply rigid statistical
methods thus excluding any philosophical approach.

        When in 1955 the senior author gave his first paper to the
Royal Statistical Society (Thom (1955)) the atmosphere was quite different.
For example, when he asked an eminent statistician how to obtain a proof
of a certain idea of his, the answer was: "It does not require proof. It

is obvious".   That particular idea has since been attacked several times.

3   LUNAR OBSERVATIONS

Let us look at one of the lunar sites, that at Brogar in Orkney, Thom & Thom (1978 a) Chapter 10.   Like many other backsights, it occupies a relatively flat area over which the observers could move to fix exact

Fig. 1.   Histogram of observed declinations, Thom (1967)

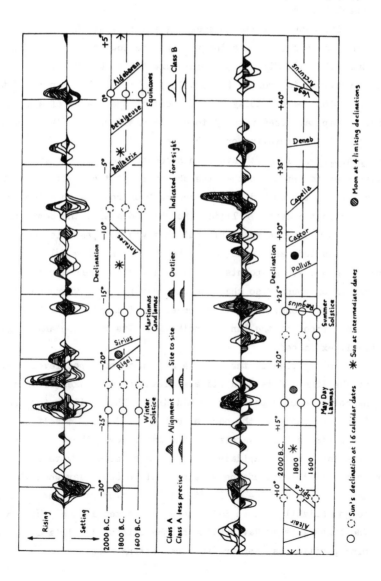

positions from which finally to observe phenomena on the horizon. The ring
itself stands on top of a gentle rise. It was built perhaps early in the
third millennium. At that time the rising Moon <u>as seen from somewhere with-</u>
<u>in the circle</u> at the minor standstill rose on the notch at Mid Hill and when
setting ran down the edge of the impressive cliffs at Hoy. As the centuries
passed the obliquity of the ecliptic ϵ fell and it was necessary to move
the sighting points outside of the south of the circle if the two foresights were to be retained. Also by this time the standard of observing
was very much higher.

The casual visitor misses a great deal at this site. It should
be noted how the old track passing through the low mound at the Comet Stone,
Thom & Thom (1978 a), Fig. 10.7, points exactly to Mid Hill. Mound L_2
should be examined in profile (from below). It will be seen that it is
practically ploughed out today. At T, most of the traces of a mound shown
on Thomas's map have been removed to enable visitors to walk round the
outer ring without stumbling on loose stones.

<u>Warning positions</u>. It must have been necessary to have warning that the
Moon was about to appear so that the observers at the Comet Stone would
be ready. The notch on Mid Hill where the Moon rose is on a long ridge
sloping down to the east. A little thought shows that the best stance for
a warner was some way to the east, but this was not possible because of the
presence of the loch and so he had to be raised and this fully explains the
presence of mound A. AB points to Mid Hill and so there would probably have
been a marker on mound B so that the warner would know where he himself
should watch. We consider that the presence of mound A is a good
philosophical proof of the lunar hypothesis. Its position and height are
so perfectly explained.

On our plan of the site we give numerous spot levels which
show that the line MLJ is on a ridge. Advantage was taken of this ridge to
place the row of mounds pointing to Mid Hill. The position for observing
Mid Hill was obviously at the base of cairn M. Cairn M itself was raised
to provide a stance for a warner for the man below. Thomas reported that
cairn M was large and dilapidated. It was also necessary to place the
warner to the north west of the site so that he could say when the Moon
was going to appear at Kame of Corrigal; but movement to the north meant
losing height slightly and so Salt Knowe was raised at the position to
suit. The foresights at Ravie Hill and the cliffs at Hellia do not need
any warning position as they are for the setting Moon.

The foresight for the Moon rising at Mid Clyth (Thom (1971)p91)
is very small. It needed a warning position. There is a row of large
boulders on the moor above Mid Clyth pointing to the notch and this was the
position to be occupied by the warner, see above. At Kintraw (Thom (1971)
p39) there is a stone above, obviously placed for a warner to let the
observers at the Susan Stone know that the Sun was coming. It is highly
probable that at many sites a row of observers stood in line and watched
as the phenomenon occurred. The position occupied by the first (or last)
to see the limb was the required point on that day.

Arguments like the above seem to us to be very important.
They cannot contribute anything to, for example, the probability level,
but without them the probability level is bare and uninteresting. The
various points made above must be used in a philosophical manner to lend
support to the astronomical argument.

4 LUNAR BANDS

To return to our present day analysis of the problem, we need
only consider those sites which give declinations falling inside the lunar
bands defined by the limits $\pm\epsilon \pm(i + \Delta + s)$ and $\pm\epsilon \pm(i - \Delta - s)$, where
ϵ is the obliquity of the ecliptic, i is the inclination of the Moon's
orbit to the ecliptic, Δ is the perturbation of the orbit and s is the
semi-diameter of the Moon. A lunar band for $\epsilon + i$ is shown diagrammatically
in Fig. 2. Relevant numerical values are given in Table 1. Eight lunar
bands exist on the horizon, four for rising and four for setting Moons.

Fig. 2. Lunar band, showing the limiting positions of the
upper and lower limbs of the rising Moon at standstill.
Appropriate numerical values of Δ and s are shown for
equinox and solstice.

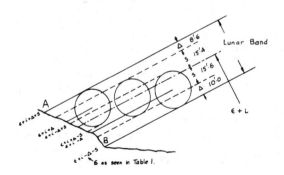

From the declinations falling within the lunar bands we deduct ($\epsilon \pm i$). If there is no astronomical significance in the foresights then there is no reason why the histogram of the points so obtained should not smear out over the lunar band widths of $2(s + \Delta)$, about 50'.0. The fact is however that they do not smear out but pile into little clumps, see Fig. 4, adapted from Thom & Thom (1980 a) Fig. 2. This argument seems to have been overlooked by many of our critics. We may have included a few spurious lines but these can not affect the main argument, more especially when this is reinforced by the philosophical points mentioned above.

4.1 Four histograms of β values

Our surveying has supplied us with declinations δ_0 of bodies which rise or set on the foresights. From these observed declinations δ_0 we deducted ($\epsilon \pm i$). Calling each difference β , where $\beta = \delta_0 -(\epsilon \pm i)$, we proceeded to compare each β with Q the "expected value". Q has one of the values of $\pm(\Delta \pm s)$.   Fig. 4 shows histograms of the values we obtained for β .

Equinoxes. Above the four histograms we have marked with wide-shafted arrows the theoretical or expected values of Q at the equinoxes. These are obtained from Fig. 3 (see Thom & Thom (1980 a), Fig. 1)which shows how Q varies with condition, major or minor standstill, and positive or negative declination. The manner in which the β values cluster round the expected, or Q, values shows that Megalithic Man was observing the declination maximum M and minimum M_1 shown in Fig. 3.

Table 1 Numerical Values of Nominal Declinations δ

	Δ	s	δ		δ
$+(\epsilon + i +\Delta)$	8.6	15.4	29° 10.4	$-(\epsilon + i +\Delta)$	-29° 10.4
$+(\epsilon + i +\Delta+ s)$	8.6	15.4	29 25.8	$-(\epsilon + i +\Delta+ s)$	-29 25.8
$+(\epsilon + i +\Delta- s)$	8.6	15.4	28 55.0	$-(\epsilon + i +\Delta- s)$	-28 55.0
$+(\epsilon + i -\Delta)$	10.0	15.6	28 51.8	$-(\epsilon + i -\Delta)$	-28 51.8
$+(\epsilon + i -\Delta+ s)$	10.0	15.6	29 07.4	$-(\epsilon + i -\Delta+ s)$	-29 07.4
$+(\epsilon + i -\Delta- s)$	10.0	15.6	28 36.2	$-(\epsilon + i -\Delta- s)$	-28 36.2
$+(\epsilon - i +\Delta)$	10.0	15.6	18 54.4	$-(\epsilon - i +\Delta)$	-18 54.4
$+(\epsilon - i +\Delta+ s)$	10.0	15.6	19 10.0	$-(\epsilon - i +\Delta+ s)$	-19 10.0
$+(\epsilon - i +\Delta- s)$	10.0	15.6	18 38.8	$-(\epsilon - i +\Delta- s)$	-18 38.8
$+(\epsilon - i -\Delta)$	8.6	15.4	18 35.8	$-(\epsilon - i -\Delta)$	-18 35.8
$+(\epsilon - i -\Delta+ s)$	8.6	15.4	18 51.2	$-(\epsilon - i -\Delta+ s)$	-18 51.2
$+(\epsilon - i -\Delta- s)$	8.6	15.4	18 20.4	$-(\epsilon - i -\Delta- s)$	-18 20.4

$$\epsilon = 23°53'.1 ; \quad i = 5°08'.7$$

Fig. 3. Conditions at the standstills.

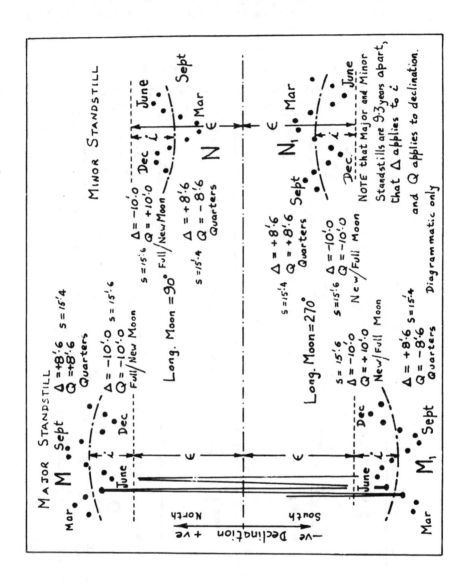

Fig. 4. Four histograms of β values, where
the expected values of β, namely Q, are shown by arrows as
in the key; Q has one of the appropriate values of $\pm(\Delta \pm s)$.

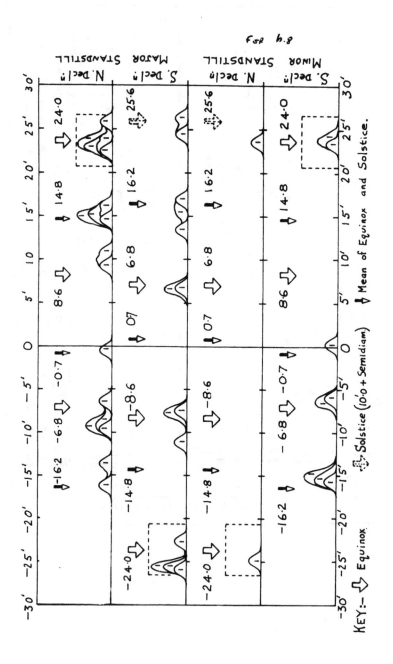

Solstices. Amongst the Q values on the histogram there are twelve values
near zero and zero ± s. These are shown in Fig. 4 by narrow-shafted arrows,
and they are placed to show the middle position between solstitial and
equinoctial values, i.e. between the declination values including + Δ
and - Δ , the top and bottom of the wobble. The Q values for these
conditions are, according to Fig. 3, ± $\frac{1}{2}$(8'.6 - 10'.0), i.e. ± 0'.7, and
to these, ± 15'.5 must be applied for the upper and lower limb (15'.5
being the mean of 15'.4 and 15'.6).

  Examination of the plotted β values on Fig. 4 shows that 14
observed lines with no Δ have Gaussians below half of the narrow-shafted
arrows (see also Thom & Thom (1980 a), Table 1). There is only one method
known to us whereby the erectors could have found these latter 14 positions
and that is by their taking a point on the ground midway between the
positions found for the top and bottom of the wobble shown on Fig. 3.
This point is midway between the position found on the ground at the standstill adjacent to the equinox and that adjacent to the solstice. A mean
of two measurements is more accurate than a single measurement and these
lines give the lowest root mean square deviation of any group we have found,
Thom & Thom (1980 a) p S84. We come to the conclusion that the erectors
were observing solstitial and equinoctial values, taking the mean; thus
they were cutting out Δ and recording (ϵ + i) or (ϵ - i), a conclusion
deduced from the observations themselves, not from any theory.

  It is interesting to see that at Temple Wood the position for
recording (ϵ + i) obtained in the above fashion is not marked by a menhir
but by a group (Q) of 4 small stones without any stone in the middle,
presumably to show that this is not an observing position, as they could
never have made accurate observations of the Moon's centre (Thom (1971)p47).

  The use that the observers made of these (ϵ ± i) positions
is explained in Thom (1971) para 8.6.

Eclipses. The eclipse limits (Thom (1971) p20) occur at the crests or
troughs of the perturbation waves at the standstills, and observations at
these times could have been used to predict the eclipses due to occur
about one fourth of a lunation later.

  4.2 Calculation of the probability level
  Clustering of β values round expected Q values. Using the
four histograms of Fig. 4 we made a rough estimate of the probability that

the β values would, if they were randomly distributed, lie as close as they do to the Q values; it turns out to be 1/10,000 (Thom & Thom (1980 a) p. S84).

We have used several times the histogram method showing "haystacks" see Thom & Thom (1978 a; 1978 b; 1979; 1980 a). These always show a very low probability level, more especially after we took the advice of L.V. Morrison from Royal Greenwich Observatory and opened the histogram out into four parts, Thom & Thom (1980 a). This is of course because of the multi-valued nature of parallax, semi-diameter and perturbation. Morrison decided the correct values to be used, see Fig. 3 and Table 1. In 1978 we thought it would improve the appeal of the method if we included only places where there was some indication at the backsight of the direction to the foresight, see the analysis in Thom & Thom (1978 b). Four or five lines in this latter paper have more than one foresight inside the lunar band, Fig. 2. This does not seriously weaken the argument. The probability level was of the order of 0.1 or 0.2%. It remains low even if four or five lines are cut out (see Thom & Thom 1978 b, p177).

Setting versus rising lines. At night the Moon was bright and the hills and sky were dark and it might be necessary to see the silhouette on the disc in order to spot the correct position, (Thom & Thom (1980), p S92). It is obviously much more easy to observe the setting rather than the rising Moon. In the list of 42 lines referred to in Thom & Thom (1980 a), Table 1, and p S85, there are 28 setting and 14 rising examples. If we discount the remote possibility that these figures are the result of chance (about 1 in 45) then we must accept the lunar hypothesis.

5 FAVOURABLE/UNFAVOURABLE METHOD

In 1979 we noticed there was another way of estimating the probability level. The method has the advantage that it is perhaps even more objective. With the approach of the major standstill, the megalithic observers would have been watching the gradual northern movement of, for example, the setting point at each lunation. The movement consisted of a series of little waves of amplitude 9' and period 173 days. It would have been natural for them to mark the peak of the highest wave at M for example in Fig. 3. In the negative case the Moon moves south along the horizon, and they would use in the same way the lowest hollow, that is the furthest south point M_1. At the minor standstill conditions were reversed

and they would be seeking and marking N and N_1. Megalithic Man would be
watching the gradual fall in the waves so he would perhaps record the
lowest, in the case of northern declination and the highest, in the case
of southern declination. With this hypothesis we decided to examine all
the lunar lines we have surveyed.

A line having a nominal declination with Δ and i of the same
sign is favourable to the hypothesis and will be marked "F". Those having
Δ and i of opposite sign are unfavourable and will be marked "U". Those
with no Δ are not listed, as they are obtained by taking the average
position of the equinoctial and solstitial observations (para 4.1).

We decided that the most objective way to choose the lines
would be to go through the two books Thom (1971) and Thom & Thom (1978 a)
page by page, and extract all values of the nominal declination of lunar
lines we found there. In all we found 51 lines, see Table 2. Recalculation
of the declinations made after the publication of the above two books,
using more sophisticated methods and techniques, have in places altered
the observed declinations slightly but this has made little change in the
arithmetical values and that only in one or two cases. Taking $\epsilon = 23°53'.1$,
(see Thom & Thom (1980 a), p S83), and the values of Δ, p and s given us
by Morrison we evaluated the nominal declinations, see Table 1. Thereafter for each value of observed declination δ_o we chose the nearest
nominal value. We also gave a class number to the lines; Class 1 when
the backsight contained some sort of indication of the direction to the
foresight, Class 2 when the indication was poor, and Class 3 when there
seemed to be absolutely no indication.

5.1 Distribution of F/U lines

Had there been a random distribution over a large number of
lines, the most probable finding would be equal numbers of F and U lines,
and yet we get 36 F's and 15 U's made up as follows: at the major standstill, 27 F's and 12 U's; at the minor standstill 9 F's and 3 U's - a
total of 36 F's and 15 U's. The calculation of the chances of this
happening by accident is the same as that required when a coin is tossed
51 times and we want to find the probability of getting at least 36 heads.
It turns out to be very small, 1 in 433 in fact.

The 15 unfavourable lines are listed in Table 3. Of these 15,
the first 8 might be called uncertain or unconvincing lines but all of
these weak lines are Class 3. If we remove from Table 2 all 18 of the

Table 2 Comparison of observed and nominal lunar declinations

Name	Site	Observed	Nominal	Diff	Nominal ±(ε ± i ±Δ ± s)	F or U	Class	Reference
Leacach an Tigh	H3/11	-29 29.5	-29 25.8	3.7	-(ε + i +Δ + s)	F	2	MLO Table 7.1 (Thom (1971))
Chloiche	H3/11	-29 04.6	-29 07.4	2.8	-(ε + i -Δ + s)	U	1	"
Callanish I	H1/1	-29 27.1	-29 25.8	1.3	-(ε + i +Δ + s)	F	1	"
(A₁)	H1/1	-29 11.6	-29 10.4	1.2	-(ε + i +Δ + s)	F	3	"
Quinish	M1/3	-29 26.9	-29 25.8	1.1	-(ε + i +Δ + s)	F	2	MLO Fig.9.2
Mid Clyth	N1/1	-29 26.6	-29 25.8	0.8	-(ε + i +Δ + s)	F	1	MLO Fig.9.3
" "	N1/1	-29 08.7	-29 07.4	1.3	-(ε + i -Δ + s)	U	2	MLO Table 7.1
" "	N1/1		-28 55.0		-(ε + i +Δ + s)	F	3	"
High Park	A4/2	-29 26.4	-29 25.8	0.6	-(ε + i +Δ + s)	F	3	MLO Fig.9.3
Small Isles	A6/4	-29 25.9	-29 25.8	0.1	-(ε + i +Δ + s)	F	3	MLO Table 7.1
Campbelltown	A4/14	-28 53.8	-28 55.0	1.2	-(ε + i +Δ - s)	F	3	"
Glen Prosen	P3/1	-28 55.0	-28 55.0	0.0	-(ε + i +Δ + s)	F	1	MLO Fig. 6.16
Callanish V	H1/5	-28 58.1	-28 55.0	3.1	-(ε + i +Δ + s)	F	3	MLO Table 7.1
" "	H1/5	-28 35.4	-28 36.2	0.8	-(ε + i -Δ + s)	U	3	"
" "	H1/5	-29 26.3	-29 36.2	0.5	-(ε + i +Δ - s)	F	3	"
Loch of Yarrows	N1/7		-28 36.2		-(ε + i +Δ + s)	U	3	MLO p.99
Camster	N1/14	-28 40.0	-28 36.2	3.8	-(ε + i +Δ + s)	U	3	MLO p.100
Brogar,Kame lower	O1/1	+29 25.4	+29 25.8	0.4	+(ε + i +Δ + s)	F	1	MRBB Table 10.1 (Thom & Thom(1978 a))
" upper	O1/1	+29 25.1	+29 25.8	0.7	+(ε + i +Δ + s)	U	1	"
" Ravie Hill	O1/1	+28 52.0	+28 51.8	0.2	+(ε + i -Δ + s)	U	1	"
Stonehenge	SS/1	-29 30±	-29 25.8	4.2	-(ε + i +Δ + s)	F	1	MRBB Table 11.2
Wormadale Hill	Z3/4	-29 08	-29 07.4	0.6	-(ε + i -Δ + s)	U	3	MRBB p.166
Ballinaby	A7/5	+29 26.6	+29 25.8	0.8	+(ε + i +Δ + s)	F	3	MRBB p.170
Beacharr	A4/5	+29 12.0	+29 10.4	1.6	+(ε + i +Δ + s)	F	2	MLO Table 7.1 (Thom (1971))
Ballymeanach A	A2/12	+29 13.6	+29 10.4	3.2	+(ε + i +Δ)	F	3	"
A2	A2/12	+29 28.5	+29 25.8	2.7	+(ε + i +Δ + s)	F	2	"
Carrach an Tarbert	A4/17	+29 26.0	+29 25.8	0.2	+(ε + i +Δ + s)	F	1	MLO Fig.6.5
	A4/17	+28 34.6	+28 36.2	1.6	+(ε + i -Δ + s)	U	3	"
The Dogs,Borgue	N1/21		+29 25.8		+(ε + i +Δ + s)	F	1	MRBB p.170 (Thom & Thom (1978 a))
Stillaig	A10/5	+28 54.2	+28 55.0	0.8	+(ε + i +Δ + s)	F	1	MLO Fig.6.11 (Thom (1971))
Temple Wood S₁	A2/8	+28 56.5	+28 55.0	1.5	+(ε + i +Δ - s)	F	1	MLO Table 7.1
Knockstaple B	A4/19	+28 54±	+28 55.0	1.0	+(ε + i +Δ - s)	F	1	MLO Table 7.2 etc.

Table 2 continued

Name	Site	Observed	Nominal	Diff.±	(ε±i±Δ±s)	F or U	Class	Reference
Camus an Stacca	A6/1	-28° 58'	-28° 55:0	3:0	-(ε+i+Δ-s)	F	3	MLO Table 7.2 etc.(Thom(1971))
" "	A6/1	-28 39	-28 37.8	1.2	-(ε+i+Δ-s)	U	3	" " "
Fowlis Wester N	P1/10	+29	+29 25.8		+(ε+i+Δ+s)	F	3	MLO p.54
" B	P1/10	+29	+29 25.8		+(ε+i+Δ+s)	F	3	MRBB Fig.13.5 (Thom & Thom (1978 a))
" "	P1/10	+28	+28 36.2		+(ε+i-Δ-s)	U	1	MLO Table 5.3 (Thom (1971))
Escart	A4/1	-28 39.3	-28 36.2	3.1	-(ε+i-Δ-s)	U	2	MLO Table 7.1 "
Crois Mhic-Aoidha	A4/9	-28 37.3	-28 36.2	1.1	-(ε+i-Δ-s)	U	3	MLO Table 7.1 "
"	A4/9	+18 18.0	+18 20.4	1.6	+(ε-i-Δ-s)	F	2	" "
Parc-y-Meirw	W9/7	+18 19.7	+18 20.4	0.7	+(ε-i-Δ+s)	F	1	" "
Haggstone Moor	G3/2	+19 09.9	+19 10.0	0.1	+(ε-i-Δ-s)	U	2	" "
"	G3/2	+18 17.8	+18 17.8	1.6	+(ε-i-Δ-s)	F	2	" "
Kintraw	A2/5	-18 20.3	-18 20.4	0.1	-(ε-i-Δ-s)	F	2	" "
"	A2/5	-18 40.6	-18 38.8	1.8	-(ε-i-Δ-s)	U	2	" "
Brogar-Hellia	O1/1	-18 50	-18 51.2	1.2	-(ε-i-Δ+s)	F	1	MRBB Table 10.1 (Thom & Thom(1978 a))
Brogar-Mid Hill	O1/1	-18 50	-18 51.2	1.2	-(ε-i-Δ+s)	F	1	" " "
Brogar-Mid Hill	O1/1	-18 19.5	-18 20.4	0.9	-(ε-i-Δ-s)	F	1	" " "
Dirlot	N1/17	+19 11	+19 10.0	1.0	+(ε-i+Δ-s)	U	3	MLO p.97 (Thom (1971))
Lund,Shetland	Z1/1	-18 17	-18 20.4	3.4	-(ε-i-Δ-s)	F	3	MRBB p.163 (Thom & Thom(1978 a))
Lundin Links	P4/1	-18 18	-18 20.4	2.4	-(ε-i-Δ-s)	F	2	MLO Fig.5.5 (Thom (1971))

Table 3 List of 15 U lines

		Name	Site	Class	
H1/5	28°35:4	Callanish V	N1/1	3	U
N1/7		Loch of Yarrows	A4/1	3	U
N1/14		Camster	G3/2	3	U
A4/9	-28 37.3	Crois Mhic Aiodha		3	U
N1/17		Dirlot		3	U
Z3/4		Wormadale Hill	P1/10	3	U
A6/1	28 39	Camus an Stacca		3	U
A4/17		Carrach an Tarbert	A2/5	3	U
H3/11	29 04.6	Leacach an Tigh Cloiche to Wiay		1	U

	Name	Site	Class	
29°08:7	Mid Clyth	N1/1	1	U
28 39.3	Escart	A4/1	2	U
	Long Tom Haggstone Moor	G3/2	2	U
	Brogar to Ravie	P1/10	2	U
	Fowlis Wester		1	U
	Kintraw	A2/5	2	U

Class 3 lines, both favourable and unfavourable, we are left with a total
of 33, made up of 26 favourable plus 7 unfavourable, and we obtain a
probability of 1 in 1500. Thus we get an improvement in probability level
when we remove the 8 weak lines and all of the 10 other remaining Class 3
lines.

It is clear that in our selection of lines as described above
we did not have anything like the F/U method in mind when we surveyed the
sites on which it is based and so no subjective bias was possible. It thus
appears that the F/U method gives definite support to the lunar hypothesis.

6 SOLAR OBSERVATION: THE CALENDAR

When we look at our collection of declinations in Fig. 1 taken
from Thom (1967), Fig. 8.1, we see that some are lunar and a number are
perhaps stellar but in addition definite clumps exist near 0°, $\pm 8^\circ$, $\pm 16^\circ$,
$\pm 22^\circ$ and $\pm 24^\circ$. The large clumps near $\pm 24^\circ$ are obviously solstitial and
that near 0° is equinoctial. It will be shown that many of the others are
also for the Sun at dates which divide the year into 16 parts that we might
call "Megalithic months". The length of the tropical year in 1800 B.C. was
365.2424 days but for the present purpose this will be approximated to $365\frac{1}{4}$
days.

6.1 The megalithic calendar in Britain

We have shown already that Megalithic Man made sophisticated
observations of the Moon. For this he must have had a calendar. A
suitable calendar has been described in Thom (1967) Chap. 9, and we
believe that this was used. Curiously, while many attacks have been made
on our lunar hypothesis, the calendar has been almost ignored. We do not
know whether this is because it is so obvious or because it appears so
complicated that it is not understood. Here it is proposed to give the
evidence existing today.

It must be emphasised that the declinations we have used were
mostly measured before we knew about the calendar and it was from these
that we deduced the calendar explained here. It had of course been
pointed out previously by Lockyer (1906) that the May Day/Lammas and the
Candlemas/Michaelmas lines at $\pm 16^\circ$ existed.

Megalithic Man could count and if he observed that the Sun rose
on a well-defined mark on a spring morning he would find that 365 days
later it rose nearly on the same mark. After a few years he would however

see that the rising point (assuming he continued to use exactly 365 days
to a year) was moving slightly to the right each year. Matters could only
be corrected if he inserted an extra day every fourth year. Megalithic Man
was certainly not inferior to modern man in intellectual ability and we
can only explain what we find by assuming that he used his ability to think
and constructed a calendar based on a 365/366 day year. With every fourth
year a leap year, his year would be $365\frac{1}{4}$ days long.

Fig. 5 shows a histogram of observed megalithic declinations
from - 25° to + 25°. Those taken from Thom (1967), Table 8.1 are listed
here in Table 4. 35 lines not in Thom (1967) and those more recently
surveyed are given in Table 5. In Tables 4 and 5 Class A refers to those
lines which have, at the backsight, some indication of the foresight; in
Class B, indication is poor or absent.

We divided the year into 16 parts (megalithic months) and
calculated the declination at the beginning of each month. For the rising
and setting cases in the histogram, Fig. 5, the declinations are shown
above and below the line respectively. It will be seen that every one of
the theoretically calculated values carries a group of observed declinations.
Broadly this is the case for the megalithic calendar.

6.2 Theoretical declinations

With 1/16 of a year for the length of a month, that is 22,
23 or 24 days, the greatest difficulty is to know which length to allot
to each of the various months. Any symmetry which our modern Julian
calendar originally had was ruined when Augustus, jealous of Julius Caesar,
left February with 28 days.

We shall call the beginning of each month the epoch. For the
present purpose let us assume that the year began at sunrise on the day
of a vernal equinox (zero epoch). On this morning the erector then set
up a backsight with a foresight for the rising Sun. Half a year later
the Sun would rise in the autumn almost on the same foresight, but 365
does not divide into two equal whole numbers. We can divide it into 182
and 183, or 183 and 182 days. In Thom (1965), para 4.2, we used 182 for
the first half year, and months of 22 and 23 days. This we shall call
Scheme A. In Thom (1967), Table 9.1, we used 183 for the first half of
the year but we allowed one of the months to have 24 days.

Here we start with 182 days for the first half of the year
(Scheme A), but retain 22 or 23 and do not anywhere use 24 days.

Table 4   List of observed lines (from M.S.B.,Table 8.1)

Class A - Definitely indicated line   Az - Azimuth
Class B - Poorer indications   h - Horizon altitude

Site		Class	Az.	h	Decl	Remarks
A1/2	Loch Nell	A	147.5	6.6	-21.8	
A1/4	Loch Seil	A	146.8	6.9	-21.3	
A2/5	Kintraw	A	223.9	0.5	-23.6	
A2/5	"	B	307.5	2.7	+21.9	
A2/12	Duncracraig	A	140.7	2.3	-23.7	
A2/14	Dunamuck So	B	138.2	3.4	-21.7	
A4/4	Ballochroy	A	315.5	0.9	+24.2	
A4/4	"	A	226	-0.1	-23.6	
A6/1	Camus an Stacca	B	213.7	4.2	-24.2	
A6/2	Strone,Jura	B	298.3	7.5	+21.6	
A6/5	Tarbert,Jura	B	106.7	1.5	- 8.1	
A8/1	Mid Sannox	A	229.3	6.2	-16.3	
A9/7	Stravannan Bay	A	136.0	2.7	-21.7	
A9/7	" "	A	311.5	0.7	+22.1	
A10/2	Lachlan Bay	A	43.0	0.6	+24.2	
A10/3	Ballimore	B	228.2	1.8	-20.6	
A11/2	Blanefield	A	56.7	+7.2	+24.0	
B1/8	Sheldon of					
	Bourtie	A	119.7	-0.2	-16.0	
B1/18	Ardlair	A	116.0	+1.1	-13.4	
B1/26	Loanhead	B	41.6	+0.7	+24.0	
B1/26	"	A	139.0	0.2	-24.3	
B2/4	Esslie (S)	A	43.1	1.1	+24.1	
B2/4	" (S)	A	306.2	0.2	+18.4	
B2/5	Esslie (N)	A	223.1	2.7	-21.2	
B3/3	Raedykes	B	314.2	2.1	+23.9	
B7/1	Clava	A	216.5	1.7	-24.3	
B7/10	Easter Delfour	A	219	2.0	-23.6	
D1/7	Barbrook	A	284.8	2.3	+10.5	
D1/7	"	B	118.6	2.2	-15.1	
G3/3	Laggangarn	B	296	2.1	+16.2	
G3/3	"	B	106.2	0.7	- 9.0	
G3/3	"	B	105.4	0.7	- 8.5	
G3/3	"	B	133.8	0.2	-23.7	
G3/12	Drumtroddan	A	43.3	0.4	+24.8	
G3/13	Wren'sEgg	B	227.5	-0.2	-23.6	
G3/17	Whithorn	B	254.3	+0.9	- 8.5	
G4/1	Carsphairn	B	100.4	3.2	- 3.5	
G4/2	The Thieves	B	228	-0.4	-23.4	
G4/12	Cambret	A	296.7	0.2	+14.7	
G4/12	"	B	116.7	5.7	-10.3	
G4/12	"	B	254.3	4.3	- 5.4	
G4/14	Cauldside	A	156.8	8.7	-23.9	
G4/14	"	B	59.5	0.3	+16.8	
G4/14	"	B	78.2	0.3	+ 6.6	
G6/2	Auldgirth	B	281.2	3.3	+ 8.9	
G7/4	Loupin Stanes	A	306.5	5.1	+24.1	
G8/8	Dere Street IV	B	276.7	2.3	+ 5.4	
G8/9	Eleven Shearers	A	94.2	4.1	+ 0.8	Exact azimuth uncertain
G8/9	" "	A	109.2	3.1	- 8.3	

Table 4 continued

Site		Class	Az	h	Decl	Remarks
G9/13	Kell Burn	A	309.8	2.9	+23.5	
H1/1	Callanish I	A	270.9	0.5	+ 0.3	
H1/1	" "	A	77.8	0.9	+ 6.9	
H1/1	" "	B	142	0.5	-24.5	
H1/3	" III	B	280	0.6	+ 5.4	
H1/3	" "	B	249	1.7	-10.2	
H1/4	" IV	B	89	1.0	+ 1.0	
H1/4	" "	B	135	1.4	-22.8	
H1/7	Gt Bernera	B	236	1.7	-16	
H1/7	"	B	83		+ 6.0	
H1/7	"	B	90	0.7	+ 0.2	
H1/10	Steinacleit	B	89.1	0.7	+ 0.7	
H1/14	Clach Stein	B	98.3	0.4	- 4.5	
H1/15	Dursainean	A	227.9	2.0	-19.3	
H2/1	Clach an Teampuill	B	138	1.9	-21.8	
H2/2	Clach Mhic Leoid	A	271.0	-0.1	0.0	
H3/1	Clach Maolrithe	A	296.6	-0.2	+13.2	
H3/2	Clach ant Sagairt	A	287.5	0.0	+ 8.8	
H3/6	Barpanan Feannag	B	220.0	0.6	-24.2	
H3/8	Na Fir Bhreige	B	288.9	2.3	+11.7	
H3/8	" " "	A	271.8	0.4	+ 0.8	
H3/9	Ben a Charra	B	255.7	-0.3	- 8.1	
H3/18	Sornach Coir Fhinn	A	303.2	0.6	+21.6	
H3/18	" " "	A	318.5	0.8	+24.0	
H4/4	Rueval Stone	A	303.8	-0.1	+16.9	
H5/1	An Carra	B	315.4	-0.1	+21.9	
H6/3	Brevig, Barra	A	135.0	-0.3	-23.6	
L1/1	Castle Rig	A	251.5	3.2	- 8.1	
L1/1	" "	A	127.0	5.2	-16.0	
L1/1	" "	A	307.0	4.6	+24.3	
L1/3	Sunkenkirk	A	128.8	0.5	-21.5	
L1/6	Burnmoor D to C	B	243.6	-0.5	-16.0	
L1/6	" D to E	B	131.9	+1.6	-21.7	Adjusted to suit average of several surveys.Az to ±6'; h to ± 2'.
L./6	Burnmoor C to D	B	63.6	4.3	+18.5	
L1/6	" C to E	B	112.3	2.6	-10.8	
L1/7	Long Meg	A	223.4	1.1	-24.2	
L1/7	" "	A	65.1	3.4	+16.7	
M2/6	Ross of Mull	A	59.9	1.5	+17.1	
M2/9	Ardlanish	B	282.4	2.6	+ 9.0	
M2/10	Uisken	B	229.6	0.3	-21.3	
M2/14	Loch Buie	A	123.4	6.8	-12.0	
M2/14	" "	A	223.6	0.4	-23.7	
M2/14	" "	A	237.0	2.1	-16.0	
M2/14	" "	B	150.8	5.1	-24.2	
N2/1	Learable Hill	A	92.8	2.4	+ 0.3	
N2/1	" "	A	61.6	2.4	+16.6	
N2/1	" "	A	75.0	2.2	+ 9.5	
N2/1	" "	B	92.8	2.4	+ 0.4	
P1/1	Muthill	A	57.3	1.8	+18.7	
P1/1	"	A	237.3	5.6	-12.7	

Table 4 continued

Site		Class	Az	h	Decl	Remarks
P1/19	Croftmoraig	B	101.7	8.9	+ 0.8	
P2/12	Dunkeld	B	310	3.7	+24	
S2/2	Merrivale	A	70.4	3.5	+14.9	
S6/1	Rollright	A	95.0	-0.2	- 3.8	
W2/1	Penmaen-Mawr	A	60.9	-0.2	+16.4	
W2/1	" "	A	240.9	+3.7	-14.1	
W5/2	Twyfos	A	118.6	5.0	-12.7	
W6/2	Rhos y Beddau	A	79.1	5.0	+10.4	
W6/2	" "	B	259.1	3.8	- 3.7	
W8/1	Rhosygelynnen	A	82.0	0.6	5.0	
W8/1	" "	A	262.0	2.2	- 3.4	Az of short line uncertain
W9/2	Gors Fawr	A	49.6	1.3	+24.3	
W9/2	" "	A	229.6	-0.2	-24.2	
W11/2	Y Pigwn	B	53.3	0.4	+21.5	
W11/2	" "	B	233.3	1.0	-21.0	
W11/2	" "	B	131.0	0.9	-23.5	
W11/4	Usk River	B	285.0	1.5	+10.1	
W11/4	" "	B	78	3.3	+ 9.7	
W11/4	" "	A	295.3	1.2	+16.0	
W11/4	" "	A	115.3	2.9	-13.1	
W8/3	Four Stones	B	67.5	0.9	+13.9	

Megalithic Man divided the year into 16 nearly equal parts, as today it
is divided into 12 nearly equal parts. The 16-month year is not superior
to the 12-month, for in neither system can the months be of equal length.

6.3 The orbit of the earth

The earth describes an elliptical orbit about the Sun, and so
the Sun appears to move around the earth in an ellipse. Particulars for
1800 B.C. are:- ϵ = obliquity of the ecliptic = $23°.906$, π = longitude
of the Sun at perigee = $218°.067$ and e = eccentricity of the ellipse =
0.0181. Longitude is measured along the ecliptic from the first point of
Aries. Let us put $\odot$ = longitude of the Sun and ℓ = longitude of the
dynamic mean Sun, that is a Sun which moves uniformly. Knowing ℓ we get
$\odot$ from

$$\odot = \ell + 2e \sin\left(\ell - \pi\right) \qquad \text{Eq. 1.}$$

Since ℓ advances uniformly with time we put the constant of proportionality
$D = 360/365\frac{1}{4}$ degrees per day and so we can write

$$\ell = Dt \qquad \text{Eq. 2.}$$

Table 5 Additional observed lines, surveyed since 1967

Site	Class	Map Ref.	Az°	h	Decl	Remarks
			127.8	9.8	-11.4	
A2/15 Torbhlarum	B	NR 863946	127.8	9.8	-11.4	"Fort" 128°
A2/16 Nr Kilmichael Glassary (Moor)	B	NR 855950	71.8	7.8	+16.4	Shaky stone. Pimple
	B	or	63.3	8.0	+21.2	Compass only. Pimple
A2/23 Craigantairbh	B	NR 862016	254.4	11.2	+0.8	Long stone.Broken off stump 71°.5
A2/18 River Add (Moor)	B	NR 933981	86.4	3.0	+3.8	8° stone orientated 86/92
A2/22 Ford (S)	B	56°10', 5°26'	215.5	5.7	-21.8	Two on old 6" O.S. map
A4/16 Culinglongart	B	NR 652120	217.3	5.9	-21.6	10° stone orientated 220°
G9/7 East Linton (E)	B	NT 617777	256.5		-6.8	10° stone 257°
H3/12 Clach Mor a Che	B	NF 770661	281.9	0.4	+6.2	Point to be used on foresight is uncertain. (Thom(1967) MSB pl30)
H7/4 Clach Ard	A	NG 420490	110.5	1.2	-10.2	Sculptured stone orientated 110°
H7/4 " "	B		46.1	2.7	+24.1	The notch is a good foresight
H8/2 Pounding Stone,Canna	B	57°03', 6°30'	118.1	3.2	-12.4	Orientated 115° or 290°
H8/2 " "	B		288.1	0.6	+9.8	Reverse
H8/4 Gramisdale Canna	B	57°04', 6°36'	65.5	0.3	+12.8	Stone 1 x 0.5 x 3.5 ft
H8/4 " "	B		245.3	1.4	-12.3	Reverse
L2/13 Oddendale	B	NY 592129	252.0	1.7	-9.1	Double circle. Outliers
N1/20 Cnoc na Maranaith	A	ND 126268	320.3	44.5	+24.1	MRBB pl71 (Thom & Thom (1978 a))
N2/3 Shin River A to B	B	NC 582049	147.5	4	-23	MRBB Table 3.7 (Thom & Thom (1978 a))
P2/8 Shianbank A to B	A	NO 156272	317.5	0.6	+24.5	Two rings. MRBB Table 3.7 (Thom & Thom (1978 a))
P2/8 " B to A			137.5	2.6	-21.9	Reverse
B3/3 Raedykes A to B	B	NO 833907	314.2	2.1	+23.9	MRBB Table 3.7. Ruinous
P1/14 Tullybeagles A to B	B	NO 010361	84	0.1	+2.9	Pair of circles. MRBB Table 3.7 (Thom & Thom(1978 a))
P1/14 " B to A	B		264	3.8	-0.3	"
W11/2 Trecastle B to A	B	SN 833310	53.3	0.1	+21.3	Pair of circles. MRBB Table 3.6
W11/14 Usk River E to W	B	SN 820258	295.3	1.2	+16.0	"
P1/10 Fowlis Wester W to E	A	NN 924250	88.5	-0.1	+0.5	Ruinous, MRBB Table 3.7.
P1/10 " " E to W	B	NN 924250	268.5	1.8	+0.3	"
P1/10 " " E to W	B	NN 924250	299.3	2 50'	+18.7	To notch. MRBB pl74
P2/17 Dowally	B	56°36'8, 3°37'8	288	4°55'	+13.7	Line of two 9' stones
P2/17 "	B		108	6°52'	-4.0	Reverse
P4/1 Lundin Links	B	403027	234	0°20'	-19.2	Foresight by map only
W9/3 Cwm Garw	B	SN 119310	259.4	5.2	-6.8	Two stones 8.5 and 7 ft to col on horizon
Z1/1 Lund	A	HY 578034	218.6	3°42'	-18.3	MRBB pl63 (Thom & Thom (1978 a))
	A		154.2	2°18'	-24.2	" "
	A		324.0	47'	+24.1	" "

where t is the number of days which have elapsed since an arbitrary zero at sunrise on the day of a vernal equinox. The declination is given by

$$\sin \delta = \sin \odot \sin \epsilon \quad . \quad . \quad . \quad . \quad . \quad . \quad \text{Eq. 3.}$$

First we want to find two days in the year half a year apart, such that the spring declination and the autumn declination are nearly equal. Using the above formulae we calculate the Sun's declination throughout the year, shown diagrammatically in Fig. 6. It will be seen that the negative lobe is shorter than the positive, that is KL is greater than LM. We draw a horizontal line $A_1 B_1$ of length a half year and find that it has a declination of about $+ \frac{1}{2}^\circ$. This is fortunate for us because when we found that we had observed a number of declinations (Fig. 5) between 0° and $+ 1^\circ$ we knew that we were on the right lines. The maximum declination in Fig. 6 is $+ 23^\circ.91$ and the minimum $- 23^\circ.91$. $A_1 B_1$ is half a year, but as we have already pointed out this has to be either 182 or 183 days.

Megalithic Man carefully divided the year into 16 months of as nearly as possible equal lengths, and so a month had either 23 or 22 days. We seek to find out how these months would be set out to get the best possible calendar. The criterion is that they must be arranged so that the declination of the first day of a month in spring, say D_1, shall be as nearly as possible equal to the declination E_1 of the first day of a month in autumn, (Fig. 6). Then each calendar foresight can serve two epochs.

6.4 Declination range throughout the 4-year leap year cycle

Let us assume, merely to illustrate, that the Sun rose on a day near the vernal equinox at $+ 0.7$ days. From equations 1, 2 and 3 we find the declination was $+ 0^\circ.79$, shown at A, Fig. 7. The sunrise at the autumnal equinox using Scheme A occurred at $182 + 0.7$ days and the declination $0^\circ.49$ is shown at B. The next vernal equinoctial rising was at 365.7 days and this gives a declination of $0^\circ.69$ shown at C. Going on like this for four years we eventually find point H. Thereafter the difference between the vernal and autumnal declinations gets bigger and bigger.

To stop this we must put in an extra day (an intercalary day) and so make one year (the leap year) have 366 days. Thus at the date 4 years + 1 day + 0.7 days = 1461.7 days, we are back at A (Fig. 7). The necessity for the extra day every fourth year must very soon have become apparent to Megalithic Man. He had set up a mark to show where the Sun

Fig. 5. Histogram of observed declinations

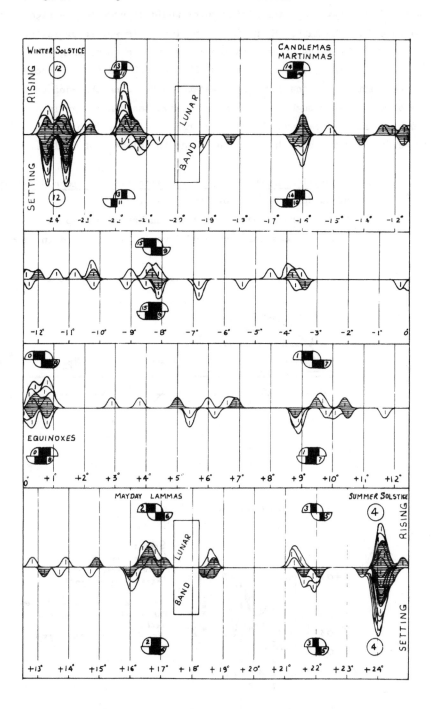

rose on the vernal equinox and so if he counted exactly 365 days in the year, after a few years the rising point would diverge to the right more and more from the mark. He must have noticed this sooner or later.

The range of declinations shown by the enveloping (dotted) rectangle in Fig. 7 is evidently $- 0°.2$ to $+ 0°.2$ and this is produced by the insertion of one day. Here we are considering the equinoxes but the other dates throughout the year will have the same range in time. No epoch in the year can ever be greater than a half day from its ideal and so we find the range in declination "r" as in column 8 of Table 6.

6.5   Calculation of the Sun's declination at each epoch.   Scheme A
The calculation is shown in Table 6.
Column 1 shows the epoch or the number of the month.
Column 2 shows the number of days we have assumed in each month.

Fig. 6. Declination of the Sun throughout the year

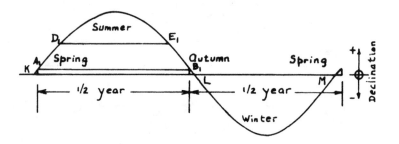

Fig. 7. Equinoctial declination

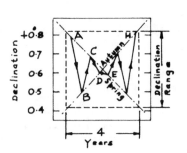

Scheme A.
Lines marked 'Spring' and 'Autumn' are independent of the starting declination.
    Declination Range indicated is the interval between the highest and the lowest declinations attained in the 4-year leap year cycle.

Column 3 is the number of days that have elapsed since zero time.

" 4 is the time of sunrise in days. We have taken this to be zero at the equinox. A month later the Sun rose about 45 minutes or 0.03 days earlier in the morning and so we subtract 0.03 from 23 and get 22.97. This kind of calculation is made for each of the 16 epochs throughout the year. We take Column 4 as t and so get ℓ from equation 2. We then get the longitude of the Sun from equation 1. The declination follows from equation 3.

Column 5 shows the values of declination, tabulated for sunrise.

Columns 6 and 7. Here we repeat the calculations for sunset. Note that in the first line at the equinox (epoch 0) the Sun set 12 hours or 0.5 of a day later than it rose. In Column 4 we had to subtract 0.03 days from 23.00 to get the earlier sunrise time at epoch 1. Correspondingly in Column 6 we have to add 0.03 days to get the later time of sunset. These allowances are made right throughout the table. Finally the declination calculated at sunset is given in Column 7.

Column 8. Angles given here show the changes in declination due to half a day's change in time. They were obtained by adding 0.5 days to t and recalculating the declination. The calculated change is called the amplitude of the declination range (r).

The values of declination we have just obtained in Columns 5 and 7 might be assumed to be at the middle of the 4 year leap-year cycle (see para 6.4). At the beginning and the end of the cycle they would be up or down by the $\pm$ r values shown in Column 8, the mean declination range.

6.6 Pairing

We now examine the values to see how successful we have been in "pairing" the spring and autumn values; for example at sunrise, at epoch 0 we get $+ 0°.52$ (see Tables 6 and 7) and at epoch 8 we get $+ 0°.77$ giving a discrepancy of $\gamma = 0°.25$. Similarly comparing epoch 1 with epoch 7 we get a discrepancy γ of $+9°.54 - 9°.18$, that is $+ 0°.36$. In this way we completed Table 7 and found finally that the "root mean square" of the discrepancy γ was $0°.24$.

Perhaps this is the place to point out that we also tabulate for the rising Sun in Column 5 the mean of the values we have paired, that is the mean of the values in Columns 2 and 3, Table 7. We also give the values for the setting Sun (Column 9) and it will be seen that the two corresponding columns are practically identical. If Megalithic Man was using this calendar then these are the values we would expect him to use, or attempt to use, but we must allow of course for semi-diameter and declination ranges (r), see Table 6, Column 8.

Table 6. Scheme A   182/183 days

Epoch Number	Days in "Month"	Epoch Nominal	Time of (days) Sunrise	δ n decl. at Sunrise	Time (days) Sunset	δ n decl. at Sunset	max. decl. range 'r'
(1)	(2)	(3)	(4)	(5)	(6)	(7)	(8)
0	23	0	0.0	+ 0.52	0.50	+ 0.71	±0.19
1	23	23	22.97	9.18	23.54	9.39	±0.18
2	23	46	45.93	16.66	46.57	16.84	±0.15
3	23	69	68.91	21.89	69.59	22.00	±0.08
4	23	92	91.90	23.90	92.59	23.91	0.00
5	23	115	114.91	22.19	115.59	22.09	0.08
6	23	138	137.93	16.98	138.57	16.79	0.14
7	22	160	159.96	9.54	160.54	9.32	0.18
8	22	182	182.00	+ 0.77	182.50	+ 0.57	0.19
9	22	204	204.3	- 8.16	204.47	- 8.33	0.19
10	23	227	227.07	-16.36	227.43	-16.47	0.15
11	23	250	250.09	-21.99	250.41	-22.04	0.08
12	23	273	273.10	-23.91	273.40	-23.91	0.00
13	23	296	296.10	-21.76	296.40	-21.71	0.09
14	23	319	319.08	-16.18	319.42	-16.08	0.15
15	23	342	342.04	- 8.38	342.46	- 8.22	0.19
16	23	365	365.00	+ 0.42	365.50	+ 0.62	0.20

The starting and finishing values in column (5) are not identical because 365 days are used in the year, and not $365\frac{1}{4}$.

Table 7 Comparison of Spring and Autumn declinations compiled from Table 6

	Sunrise				Sunset			
Epochs (1)	Spring (2)	Autumn (3)	Discr(δ) (4)	Mean (5)	Spring (6)	Autumn (7)	Discr(δ) (8)	Mean (9)
✳ 0 and 8	0.52	0.77	+0.25	0.64	0.71	0.57	-0.14	0.64
1 " 7	9.18	9.54	+0.36	9.36	9.39	9.32	-0.07	9.34
2 " 6	16.66	16.98	+0.32	16.82	16.84	16.79	-0.05	16.82
3 " 5	21.89	22.19	+0.30	22.04	22.00	22.09	+0.09	22.04
✳16 " 8	0.42	0.77	+0.35	0.59	+ 0.62	+ 0.57	-0.05	+ 0.60
15 " 9	- 8.38	- 8.16	+0.22	— 8.27	- 8.22	- 8.33	-0.11	- 8.27
14 " 10	-16.18	-16.36	-0.18	-16.27	-16.08	-16.47	-0.39	-16.27
13 " 11	-21.76	-21.99	-0.23	-21.87	-21.71	-22.04	-0.33	-21.87

✳ These two values are not identical because 365 days were used in the year, not $365\frac{1}{4}$.

R.M.S. $\delta = 0°.24$

7   HISTOGRAM OF OBSERVED DECLINATIONS

The declinations available up to 1967 are shown in Thom (1967)
Table 8.1. Some of these values must now be dropped, e.g. lines like those
taken looking outwards from Temple Wood to the stones in the meadow, because
as has been shown since, these stones are mostly backsights for lunar lines,
see Thom (1971).

Every card in our files was examined and reconsidered afresh.
Any lines with doubtful azimuth or altitude were omitted. We found that
we were throwing out some of those we had previously put in the lowest
category C. We have tried to be objective in selecting the sites to be
included and to avoid being influenced by a knowledge of the important
declinations.

We give in Table 4, the lines obtained from Thom (1967) Table
8.1 and in Table 5, lines which do not appear in that book (see para 6.1).
Amongst these are lines from sites which have been surveyed since Thom
(1967) was written. As some of these lines in Table 5 have not been published before we now list them all with map coordinates, azimuths, altitudes
and deduced declinations. Since no calendar declination can be greater
than that of the Sun at the solstice we have limited the histogram to the
range $- 25^{\circ}$ to $+ 25^{\circ}$, but as this range includes the declination of the
Moon's minor standstills we have marked on the histogram the "lunar bands"
at $+ 17^{\circ}.4$ to $+ 18^{\circ}.2$, and $- 19^{\circ}.3$ to $- 20^{\circ}.1$. These bands cover all
possible positions of the Moon's upper and lower limb in the sky at the
minor standstill.

8   SOLSTITIAL VALUES

At the winter solstice in some sites the erectors used the upper
limb and in some the lower. The double peak on the histogram shows this
very clearly. At the summer solstice they seem to have used mostly the
upper limb. Dr. E.W. MacKie pointed out that at the summer solstice the
Sun would be so bright that it would often not be possible to make use of
the lower limb. At the winter solstice in these latitudes it was probably
possible to use either the upper or the lower limb without much trouble.
Perhaps they used something equivalent to a dark glass.

We are sure of the calendar declination only at the winter and
summer solstice. For the other epochs the erectors may have used Scheme A
as above or some other similar scheme, for example 183 and 182 days for the
first and second half year respectively. A scheme with 183 and 182 days

is shown in Thom (1967) p110, Table 9.1. It gave a lower root mean square
residual,but it had one month with 24 days. We examined a type B scheme
with 22/23 day months and repeated the whole calculation but it showed that
Scheme A was superior. There may be other schemes but before we can
really decide mathematically or statistically which scheme was used it
will be necessary to have more observations. There are still plenty of
sites throughout Britain and France which have not been measured up and
when our measurements have been checked and these extra sites have been
surveyed it should be possible to form the difference between the observed
and the calculated declinations for each line, and from the mean, to form
a better opinion as to what scheme was used.

8.1   Scheme A on the histogram

On the histogram we show for Scheme A the theoretical declinations
using small figures similar to those in Fig. 8. Each is different in
dimensions (see Tables 6 and 7) but all are like Fig. 8 where the mean
declination is shown at 00. The declination range r from Column 8 Table 6
is shown in Fig. 5 by black rectangles. The lack of pairing is shown by
the distances (γ) between the centre lines of the "spring" and "autumn"
rectangles and the Sun's semi-diameter is shown by the arcs at the ends
of the black range rectangles. At the solstices, r and γ have shrunk
to zero and the figure becomes simply a complete disc representing the
Sun. In Table 7 it will be noticed that the mean sunset and sunrise
declinations are equal. This also appears in the little figures superimposed in Fig. 5, although the range (r) and the discrepancy (γ) are
different for each. Since we start numbering at the equinox (No. 0) and
finish with No. 16 at the equinox there is an extra pair of diagrams here
but for clarity we have not plotted these.

It should be remembered that all the values shown on the histogram are liable to inexactitude due to errors in our surveying, errors due
to the stones having been moved or even errors made by the original
erectors. The first mentioned source of error can of course be removed
by an independent observer making careful new measurements.

It is not necessary to understand all the above, especially
paras 6.3 to 6.6, and para 8.1. What is important is to note that
Megalithic Man had a solar calendar in which the declinations must have been
very close to those listed.

9   DATING BY ASTRONOMICAL METHOD

Lunar lines. We can only assume that the lunar observatories
were all erected at the same time. In Thom & Thom (1980 a) pp S81 and
S83 it will be seen that we obtained the average value of $23°53'.1$ for ϵ
from the 42 lines, and this yields a date of about 1590 B.C. $\pm$ 100 years.
A possibility of error is introduced here because all of the 42 lines were
probably not erected at the same time, but we draw attention to the small
value (1'.34) of r, the root mean square of R where R $= \beta$ - Q. In this
connection 1'.34 is equivalent to about 200 years.

Solar lines. Obviously exact solar dating can be attempted only at the
solstices. Three first class lines have been discovered, Thom (1971) p 44,
yielding the date of 1750 B.C. ($\pm$ 100 years).

Broadly speaking therefore the more precise solar and lunar
lines are contemporary.

10   REFRACTION

An examination of the very full tables of refraction published
by the Nautical Almanac shows that, provided the observed altitude is not
negative, there is not much trouble about refraction. We draw attention

Fig. 8. Conventional representation of declination at calendar
dates as shown in Fig. 5

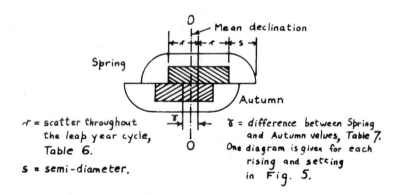

to work done on refraction measurement (Thom 1958). At very low and
negative altitudes the refraction becomes uncertain but only a few lines
like this exist.

A much more serious problem is that of graze. We have shown
from our observations that the mean graze is about one minute and it is
interesting to read the earlier analysis in Thom (1969) p 21 where it was
shown that the astronomical refraction which we were then using should
be increased by 2 or 3%. Graze of course was the cause: at that time
we were using no correction for graze but we use it now.

11 CLIMATE

It is known that there was a spell of dry climate around 2,000
B.C., see Evans (1975), pp 142-145, but there does not seem to be any
definite idea how dry it really was. We have elsewhere shown that the
original observers could have replaced, on the ground, a single observation
missed due to cloudy weather. We obtain no dates later than the middle of
the 2nd millennium and meteorologists consider that the weather began to
deteriorate then.

12 SPURIOUS LINES

Several critics have said that our work includes a number of
spurious lines and that is how we have got the favourable lunar results.
We have here shown earlier, by a consideration of the observations confined
to the lunar band, that the above criticism is wrong because when we made
each measurement in the field we could not have known the exact declinations
and so could make no subjective decision. Similarly no subjective decisions
were made either to include or exclude Megalithic Man's 13 equinoctial
lines with declination about $+ \frac{1}{2}^{\circ}$ (Fig. 6 para 6.3).

13 CONCLUSION

We must make it clear that although we have used a modicum of
mathematics in studying the remains, Megalithic Man did not need to know
these mathematics. He probably had his own method of reasoning.

Lunar hypothesis. There are many more sites in Britain still to be surveyed.
When information from these is available it may be possible to obtain a
better idea of how Megalithic Man worked. Examination of the declinations
within the lunar bands shows a histogram which is definitely favourable

to our hypothesis. The Favourable/Unfavourable method also gives strong
support, more especially as nothing was known of this method when the
surveys upon which it is based were made and published. In the Favourable/
Unfavourable method of presentation of data we get a much better result
when we remove the class 3 lines. At a number of lunar sites we find
positions prepared for warners to occupy. We might add that at Stonehenge
the warner probably occupied the top of one of the lintels.

It is true that the histograms inside the lunar bands consist
of a series of humps about 9' apart, which occur near the expected values,
Q, and so, with the observed declinations, we have a choice into which
position to put the resulting values of β . Had we tried to move from
the positions actually used on the histogram, we would have had to go back
or forward about 1,300 years to make things fit, a totally impossible
span of time.

The number of lunar setting lines is significantly more than
the rising lines, para 4.3, a point which strengthens the case that we
are dealing with intended positions, since it is easier to observe the
setting than the rising Moon.

Solar calendar. We find on the histogram a clump of declinations near to
+ $\frac{1}{2}^{0}$ and this is the declination to be expected if the year was equally
divided, Megalithic Man's only way of defining equinox. It seems certain
that Megalithic Man was carefully observing solstitial rising and setting
points and it is highly probable that he did use a 16 month calendar based
on a year of 365/366 days. We have produced a theoretical calendar with
months (epochs) of 22 or 23 days but we can not be sure that this is
exactly what the erectors used. At the summer solstice the upper limb
was observed, at the winter solstice both limbs. At epochs 10 and 11 we
have evidence for the use of the upper limb only but elsewhere things
seem rather mixed.

Taking all of the above points together we have no doubt that
Megalithic Man was making use of astronomical methods. Other than for
reasons connected with husbandry we can only theorise why such an advanced
calendar might have been devised. As the date of the calendar lines appears
to be about the same as the date of the lunar lines, early in the second
millennium, we believe that the solar calendar might have been used to
advantage in the study and prediction of the movement of the Moon, especially
at the standstills. At these periods the observers, using only their method

of measuring azimuth on the horizon, had their only chance of latching on
to the system and predicting eclipses.

Was the motive curiosity? Why was so much energy expanded on
erecting the stones? Did they want a better understanding of their
environment?

From information obtained in the field we put forward reasons
which fit the alignments formed by many of the stone rings, lines and large
single slabs.

Acknowledgment.   The authors thank Professor A.E. Roy of the Department of
Astronomy, The University of Glasgow, for reading through the manuscript.

References.
Evans, J.G. (1975). The environment of early man in the British Isles.
         Paul Elek.
Lockyer, Sir N. (1906).   Stonehenge and other British Stone Monuments.
         Macmillan.
Thom, A. (1954). Some refraction measurements at low altitude.
         Jour. Inst. Navigation, 7(3), 301-304.
Thom, A. (1955). A statistical examination of the megalithic sites in
         Britain. Journal of the Royal Statistical Society. A.118,275-295.
Thom, A. (1958). An empirical investigation of atmospheric refraction.
         Empire Survey Review, 14, 248-262.
Thom, A. (1965). Megalithic astronomy: indications in standing stones.
         In Vistas in Astronomy, ed. A. Beer, Vol. 7, pp 1-57. Pergamon.
Thom, A. (1967). Megalithic Sites in Britain, Oxford. Clarendon.
Thom, A, (1969). The Lunar Observatories of Megalithic Man.
         Vistas in Astronomy, ed. A. Beer, Vol. 11, pp1-29, Pergamon.
Thom, A, (1971). Megalithic Lunar Observatories, Oxford. Clarendon.
Thom, A. & Thom, A.S. (1978 a) Megalithic Remains in Britain and Brittany.
         Oxford. Clarendon.
Thom, A. & Thom A.S. (1978 b) A reconsideration of the Lunar Sites in
         Britain. Journal for the History of Astronomy, ix (1978), 170-179.
Thom, A. & Thom, A.S. (1979). The Standing Stones in Argyllshire.
         Glasgow Arch. Journal, Vol. 6 (1979), 5-10.
Thom, A & Thom, A.S. (1980 a). A study of all Lunar Sightlines.
         Archaeoastronomy.2. Suppl. to J.H.A. Vol. xi, 1980, S78-S89.
Thom, A. & Thom, A.S. (1980 b) Astronomical Foresights used by Megalithic
         Man. Archaeoastronomy.2.Suppl. to J.H.A. Vol.xi,1980, S90-S94.
Thom, A.S. (1981). Megalithic Lunar Observatories: An assessment of 42
         Lunar Alignments. In Astronomy and Society in Britain during
         the period 4000 - 1500 B.C., ed. C.L.N. Ruggles & A.W.R. Whittle,
         pp 13-61, BAR British Series 88, 1981.

MEGALITHIC ASTRONOMICAL SIGHTLINES: CURRENT
REASSESSMENT AND FUTURE DIRECTIONS

C.L.N. Ruggles
Dept. of Archaeology, University College, P.O. Box 78,
Cardiff CF1 1XL, Wales

Abstract. A general account and progress report is
presented concerning the aims of, and conclusions drawn
from, two extensive seasons of fieldwork on Scottish
megalithic sites. Topics discussed fall into three main
categories: reassessments of existing work by A. Thom &
A.S. Thom; suggestions for methodological improvements;
and fresh site investigations carried out according to a
more explicit methodology. It is concluded that the case
in favour of high precision astronomical sightlines, as a
primary function of many sites, is unproven on the basis
of the evidence currently available. The suggestion is
made that future work in British archaeoastronomy should
concentrate on rougher astronomical alignments and
indications, regarding them as merely one of any number
of factors that might have given rise to any particular
structure orientation. It is also recommended that more
attention should be focussed upon groups of sites which
are demonstrably similar archaeologically, where marked
orientation trends are evident at the outset, rather than
upon archaeologically somewhat diverse collections of
sites, which has often been the case up to the present.

INTRODUCTION

This paper consists of a general progress report rather
than a presentation of results: various detailed accounts have been
and are being published elsewhere (Ruggles 1981, 1982, 1983). It is
written upon the completion of a $3\frac{1}{2}$-month continuous period of
surveying work in Scotland during May - August 1981, before any
processing of the raw field data has even begun. Its purpose is to
try to explain the general aims which have motivated the fieldwork
to date, and to present the conclusions so far drawn. The 1981 season
followed on from a similar trip over the summer of 1979; some of the
results from that trip are now in print (Ruggles 1981), but much of
the 1979 field data also has still to be reduced. This paper consists
largely of generalities about procedure, rather than a succession of
remarks about individual sites.

The basic aims of the author throughout the project have been the following:

(1) To reassess existing site data. This means almost exclusively reassessing the work of A. Thom (1967a, 1971) and A. Thom & A.S. Thom (1978a,b, 1980; A.S. Thom 1981). In megalithic astronomy, any attempt to make progress inevitably has to begin with a critique of the Thoms' procedures and conclusions, simply because they have done such extensive and high-quality groundwork, and it is upon this that we need to build.

(2) To use the insights gained from the reassessment, in order to suggest improvements in the methodology of site work. Examples of methodological considerations are the way an investigator might go about choosing which particular alignments to consider at a given site as possible astronomical indicators, or which horizon features to consider as candidates for astronomical foresights. This sort of consideration is actually crucial in determining what results he might come up with. These points will be discussed in more detail in what follows.

(3) To carry out fresh site work according to an improved methodology, in the hope of producing new results which are of value.

REASSESSMENTS OF EXISTING SITE DATA
Levels of analysis by the Thoms

The evidence in favour of megalithic astronomy in general, and particularly the lunar sightlines, on which the first part of this paper will concentrate, is cumulative. At each stage in the Thoms' work over the past 25 years, the essential evidence consists of one or more analyses of many putative sightlines from a number of megalithic sites taken together. At each stage the claimed precision of the sightlines that emerges from the analysis is greater than was claimed at the previous stages. In other words, the hypothesis of high-precision lunar sightlines essentially rests upon a succession of broad-based analyses testing for ever greater levels of precision in the sightlines. The Levels of analysis, as identified in a previous paper (Ruggles 1981), are as follows:

Level 1. The analysis of over 250 possible sightlines at 145
sites (Thom 1967a, table 8.1 & fig. 8.1). Most of these
sightlines are defined by indications on the ground such as
alignments of stones. The results of Thom's analysis are
displayed in the form of a sort of probability histogram, or
"curvigram" (Ruggles 1981, p.156). It shows accumulations of
probability at various declinations, including the four lunar
extreme declinations referred to by Thom as the "standstills"
(Thom 1971, p.18). Because of the accuracy to which each
indication is measured, in other words the "spread" of each
probability hump, the accumulations of humps suggest the
existence of lunar sightlines of a rough nature, set up to a
precision of about half a degree. (The possibility that they
may be more precise has not, of course, been excluded.)
Level 2. If we superimpose the four declination intervals
centred upon the lunar extremes, we obtain another curvigram
given by Thom in his first book (Thom 1967a, fig. 10.1). If
half a degree was the best precision that was achieved, we
would expect something like a normal distribution curve to
appear, with its peak at the centre of the graph. In fact,
however, the data is suggestive of a bimodal distribution,
with peaks at around 0.25 degs, either side of the centre.
This suggests that the two limbs of the moon were preferentially observed, increasing the inferred precision to about
10 minutes of arc. The formal difference between the Level 1
and Level 2 analyses here is that in the first case we are
testing for a unimodal (Normal) fit to the curve, and in the
second for a bimodal fit; though in both cases we are testing
the same data.
Level 3. In later work Thom tested a different hypothesis on
the ground, namely that distant horizon features such as
notches provided natural foresights, and that structures at
the sites themselves simply identified the observing position
and which foresight was to be used. Because of the accuracy
of any individual distant foresight, the spread on each gaussian hump is now much smaller than before, only 3 or 4 minutes
of arc instead of something like 25. In his curvigram at
Level 3 (Thom 1971, fig. 7.1) Thom plotted the modulus of the

difference from the mean lunar extremes, and there is clear
evidence of probability accumulations around values of s,
s-Δ and s+Δ , especially the last of these.

Level 4. The analysis at Level 3 took no account of certain
variable small corrections to the declinations, such as variable parallax. At Level 4, represented in three papers by the
Thoms since 1978 (Thom & Thom 1978b, 1980; A.S. Thom 1981),
each sightline is considered on its own merits, taking into
account the time of year and the time of day of presumed use,
and various small corrections are made accordingly. In addition, the data set has been restricted to those few lines
the Thoms consider to be most convincing on the ground. The
resulting curvigram peaks are even more clear cut than at the
previous levels, and when the analysis is reformulated in a
more rigorous way as suggested by Morrison (1980), the sightlines seem so accurate (to better than a minute of arc) that
it seems they could only have been set up at the end of an
averaging process lasting some 180 years, a point acknowledged
by A.S. Thom (1981, p.38).

General Remarks

The central question in reassessing the various analyses of
the Thoms is whether each given set of putative sightlines could quite
adequately be explained away as chance occurrences. As pointed out by
Heggie (1981a, pp.S18-9) assessing the formal statistical significance of
any results is relatively easy. The major problem is to satisfy oneself
that the sightlines chosen for analysis have been selected fairly, that is
in a manner totally uninfluenced by the astronomical possibilities. This
is not in any way meant to imply that the Thoms might have been deliberately misleading us all along by carefully choosing only those lines which
best fit the theories they are trying to prove. But we do have closely
to investigate the Thoms' implicit methodology. At every stage in the
process of investigating, measuring and interpreting a site, procedural
considerations can, and in the absence of an explicit methodology might
very considerably, introduce subjective bias into the final data set upon
which the investigator performs his statistical tests. Examples are the
way one decides which sites to visit and survey in the first place, the

way one decides which indications to look at and survey at each site when
actually visiting it, and the way one decides which horizon features are
acceptable as potential astronomical foresights and can be included in
the final analysis. One of the author's principal interests has been to
investigate the selection problem at each of the Levels, and to consider
ways of improving each stage of the investigative procedure in order to
eliminate as far as possible the likelihood of selective bias affecting
the final data set. The first stage in this investigation is to try to
reassess the Thoms' own methodology at each of the Levels.

Levels 3 and 4 - the precise lunar foresights discussed by the
Thoms in recent years - are considered first because here the author's
conclusions are most negative, and lead to the assertion that we shall be
safer concentrating our future attention on other areas, at least for the
time being. A detailed account of the Level 3 reassessment has already
been given (Ruggles 1981, section 4.4), so only the salient points will
be summarised here.

Reassessment of Level 3

The analysis by Thom of 40 sightlines (1971, table 7.1) yields
a graph virtually identical to the one replotted in fig. 1a. The peaks
correspond well to deviations of s, s-Δ and s+Δ from the mean standstills, especially s+Δ at around 25'. That this is so should in fact
already raise our suspicions, for even if the wobble extremes were observed, the 190-year cyclical variation in these declinations, due to the
variation in the parallax correction, should in fact fuzz these points
out over a range about 6' wide. The reader will recall that a fixed parallax correction was applied to this data by the Thoms at Level 3. When
the 40 sightlines are visited, it transpires in fact that 1 foresight is
non-existent and a further 5 cannot actually be seen at all from the
position from which they are meant to be indicated - an example being the
line to the south at Callanish (Ruggles 1982, line 14). The foresight,
amongst mountains on Harris some 26 km away, is obscured by a local outcrop just 50m to the south. A further 19 of the 40 claimed foresights are
not in fact indicated on the ground at all (or else the supposed indication is not genuine or is some degrees off line). This fact is not clear
from reading Thom's accounts (1971, chs. 5 & 6). A typical example is
Beacharr in Kintyre (Thom 1971, pp.60-1), a 5m-tall menhir with its longer

sides oriented roughly parallel to the shore; the claimed foresight, however, is out to sea on Jura, where there is a fine selection of notches
for the investigator to choose from as putative foresights - without
restriction, since none is actually indicated.

Now we could, of course, choose to test the hypothesis that
there exist astronomical foresights at a site that weren't actually
indicated, or were indicated by markers that have disappeared. But if we
do that, we can not, by the very nature of the hypothesis, tell which of
the hundreds of notches typically on the horizon at any given site was
actually used as a foresight. Only a handful could have been so used,
even after the most exceptional site placement. In order to test the
hypothesis we would have to survey and analyse all the notches at each
site; and only by analysing probably hundreds of sites in this way could
we hope that any significant declinations would begin to show above the
random noise, even if all the sites did have one or two notches used as
foresights. Obviously a surveying project like this would be the most
monumental waste of time. On the other hand if we simply went round
surveying unindicated notches in astronomically "interesting" directions,
we would merely be fitting an unique theory to each site and actually
proving nothing. In other words, even if unindicated notches were used,
it would be a hopeless task trying to prove it. To put this point another
way, the possibility of obtaining evidence for the observation of unindicated foresights is one made very difficult, if not completely impossible,
by the limitations of the archaeoastronomical record. Our only hope is
to stick to indicated foresights.

Returning to Thom's Level 3 analysis, in fig. 1b we have
omitted the non-existent foresight and the five which can not be seen from
the indication, and the unindicated foresights are left unshaded. It is
clear that the main peak at $s+\Delta$ is largely formed by unindicated foresights. The fact that a peak shows up at all amongst so few unindicated
lines, let alone so prominently, is tantamount to proof that Thom has
simply included in his analysis those horizon features that fitted the
anticipated declinations in the first place. This, in the author's view,
simply backs up the variable parallax argument for thinking the same
thing.

Some specific examples of how selective bias comes about have
been given previously (Ruggles 1981, p.182). It is clear from these

Fig. 1. Reassessment at Level 3, after Ruggles (1981).
(a) The analysis of 40 sightlines at Level 3, as presented by
Thom. (b) Modified analysis, with six rejected sightlines
excluded (see text) and 19 representing unindicated foresights
left unshaded. The two humps marked "●" should have been
shaded (see Ruggles 1981, p. 198, note 1). (c) An independent
analysis by the author of 13 of the remaining 15 indications,
using a more explicit selection procedure for putative foresights (see text).

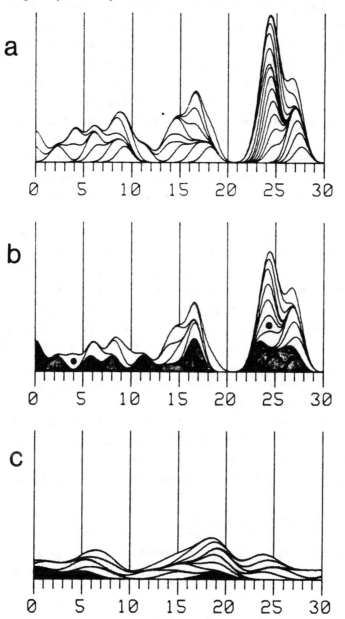

examples that horizon features were selected for inclusion in the Level 3
analysis on the basis of whether their declinations corresponded to a
plausible deviation (in terms of the lunar perturbation theory) from a
mean lunar extreme. Patently the results of any statistical analysis upon
this data set are meaningless because it has been selected to fit the
hypothesis in the first place. It remains to ask how this has actually
come about. The reason can be found amongst recently published accounts
by the Thoms of their site procedure (A.S. Thom 1981, p.24). It is clear
that if a horizon feature is found, by site measurement, to yield an
astronomically significant declination, then it is considered by the Thoms
as a candidate for a deliberate foresight, whereas if it does not, it is
not so considered. The Thoms feel quite justified, in view of the pre-
existing conclusions at Levels 1 and 2 when this later work was carried
out, in proceeding in this manner. However the author submits that this
approach merely results in a data set heavily favouring an astronomical
interpretation, and that the case in favour of precise lunar sightlines
at Level 3 is in fact unproven on the basis of the data presented by
Thom.

Of the 15 remaining indicated foresights in the Level 3 analysis, it transpires that the types of horizon foresight are not in fact all
notches, but are various: there are notches, sides of hillslopes, knobs,
dips, and tops and shoulders of hills (Ruggles 1981, p.186). In all but
three cases, within each range of horizon that might have been indicated,
there are at least one or two, and often several, other contenders for
indicated foresights, equally acceptable on the ground but which were not
included in the analysis because their declinations did not correspond to
any of the lunar perturbation extremes considered by Thom. Fig. 1c shows
the results of an independent analysis by the author, details of which
have been given previously (Ruggles 1981, section 4.4.4), which includes
only notches and dips - but all of them - that could equally well have
been indicated by each relevant suggested indication on the ground used
by Thom. It is clear that no significant peaks are now evident.

Reassessment of Level 4

At Level 4, variable parallax is taken into account by the
Thoms (Thom & Thom 1978a, ch. 2); and indeed at Brogar, for example, they
actually use the variation to calculate a date of use for the site -

approximately 1700 BC (Thom & Thom 1978a, ch. 10). It is the author's
contention that there is a significant selection effect present at Level 4
also; one which explains the small - really anomalously small - statistical residuals (Thom & Thom 1980, pp.S87-8). It enters because of the
"nominal values" for the declination assigned to each sightline. What
happens is that each sightline is weighed up by the Thoms in advance of
the analysis, and its function determined - which of the lunar perturbat-
ion values $\pm$ ($\epsilon \pm$ i $\pm$ s $\pm \Delta$) it represents. There is also a considerable
degree of freedom in choosing the observing position for each sightline,
since the indications are wide and some of the observing positions are not
actually behind one of the putative indications. It is probably impossible
to demonstrate directly that chance occurrence and selection can account
for all of the Thoms' results at Brogar, but one project in progress is
to use independent survey data (obtained in 1979) and to start out with
the supposition of a rough date of use wildly different from that assumed
by the Thoms, in other words to have in mind at the outset a significantly
different value for the obliquity of the ecliptic, ϵ. A date of 3000 BC
would probably be sufficiently different from 1500 BC for this purpose,
although it might be better to use 4000 or 5000 BC simply to take the new
supposed date totally beyond the bounds of archaeological possibility.
One then proceeds to assign different nominal declinations to the foresights, deciding independently on observing positions, and tries to produce as many putative sightlines at the site as possible, a procedure in
accord with that of the Thoms at Brogar. It is important to note that if
several "sightlines" can be produced in this way, then any precise date
of use calculated back from the new nominal functions of the foresights
and associated observing positions will now be near to the new rough date
of use assumed at the outset. The results of the completed analysis will
be published shortly (Ruggles 1983).

At Level 4 the Thoms have selected only those sightlines which
they consider suffer least from various objections (such as those raised
here at Level 3): thus in one analysis (Thom & Thom 1978b) they omit
unindicated foresights. The author has now resurveyed all but two of the
latest 42 sightlines, and one of the projects in progress is to complete
a thorough reassessment of Level 4 on the basis of this data. A critique
of the backsights, indicators and the archaeological status of the sightlines is given in a paper currently in press (Ruggles 1982) and foresights

and the problem of selection will be discussed in a paper to follow
(Ruggles 1983). However it would be misleading not to admit that the
expected results of this reassessment are just as negative as at Level 3.
Details will be given in the two papers referred to above.

In conclusion about the lunar foresights of great precision,
the author has found no evidence to suggest that the Thoms' putative sightlines cannot be completely explained away as chance occurrences, given the
selection that has gone into the alleged foresights; however a final judgement must await the outcome of the completed Level 4 reassessment. While
we should of course keep an open mind as to the existence of precise
astronomical foresights, it seems that for the time being at least our
attention is best devoted to the rougher indications discussed by A. Thom
in his earlier work, i.e. to Levels 2 and 1.

Reassessment of Level 2

This discussion will be brief, as a full account has been given
previously (Ruggles 1981, section 4.3). As stated earlier, it is the
bimodal shape of the curvigram (Thom 1967a, fig. 10.1) which is significant
at Level 2. Thom has stated (1967a, p.165; 1967b, pp.95-6) that this shape
came as a complete surprise when he plotted the graph. After reassessment of all the sightlines, based on survey work during 1979 and before, the
author concluded that no significant evidence seemed to remain for a
bimodal, as opposed to a merely unimodal (or Normal) distribution. As
there seems to be no way that the Thoms could have produced the bimodal
shape by preferential selection of lines, if the author's reassessment is
accepted we have to assume that the bimodal shape came about by chance.
Heggie (1981b, pp.243-4) has shown that a unimodal distribution in fact
provides a satisfactory fit to Thom's data. In view of these considerations it seems that we should accept the "chance" explanation for the time
being. As soon as we can achieve a satisfactory reassessment of the data
at Level 1, we can return to Level 2.

Reassessment of Level 1

Upon the Level 1 analyses of Thom (1955; 1967a, ch. 8) rest
not only the lunar sightlines but also the solstitial and calendrical ones;
in fact, these analyses provide the essential evidential basis for the
whole of megalithic astronomy above the level of the roughest orientations

of sites. At Level 1 Thom simply plots the declinations indicated by a
variety of structures on the ground - alignments of stones, circle to outlier indications, flat sides of slabs and so on- at a selection of megalithic sites.

     The first question we must ask is how the sites were chosen
for inclusion in the lists. The earliest curvigram (Thom 1955, fig. 8)
is clearly stated to contain "all the available data", in other words all
the sites surveyed up until then. However when we reach the later and
much more extensive curvigram (Thom 1967a, fig. 8.1), no such assurance
is given. The author has pointed out (Ruggles 1981, p.156) that the
sequence of site reference numbers in Thom's reference site list (Thom
1967a, table 12.1) - A1/1, A1/2 etc. - shows that for each site included
in this list, something like two other sites, which were presumably
present in some unpublished full site list, were missed out. We must ask
why all these other sites were found unsuitable for surveying and for
inclusion in the book. "Legitimate" reasons, if it is appropriate to use
the term, would be ones that would not introduce any overall bias into
the results, in other words would be reasons unrelated to the possible
astronomical significance of the site. Examples could be that it was
raining when the investigator visited the site and so he could not survey,
or the site was not a genuine megalithic one or was in too bad a state of
repair. If on the other hand, for example, the investigator visits a site
armed with a magnetic compass, and looks to see if there are any indications in astronomically "interesting" directions, and then on the basis of
that decides whether or not to stay there and survey, then his eventual
results will come out biased towards supporting whatever astronomical
hypotheses were considered "interesting". Since that uncertainty was
pointed out, the Thoms have been kind enough to send the author a Xerox
copy of their unpublished full site list. What emerges from studying that
list is that the bulk of the sites missed out of "Megalithic Sites in
Britain" were antiquities marked on the old 6" Ordnance Survey maps of
around the turn of the century, and many of them are non-megalithic, of
uncertain archaeological status, or else in a thoroughly hopeless state
of repair. Thom missed out from his full site list many megalithic sites
equal in archaeological status to the ones he included, but the bulk of
these sites were not shown on the old maps that he used, and so his reasons
for missing the sites were demonstrably largely independent of any astronomical considerations and so quite "legitimate". In other words, nothing

was found in the selection of the sites themselves which would have significantly influenced the results of the astronomical analysis.

The second question is a much more tricky one. This concerns
the selection of potential indications at each site. In the 1955 analysis
rigid selection criteria were adhered to by Thom. He considered only
indications defined by an outlier to a circle, two slabs in line, or a row
of three or more stones. It is worth noting that even in this early
analysis accumulations of humps are evident at the solar solstices and
equinoxes. In the later analysis (Thom 1967a, ch. 8) however, there are
no such clear-cut selection criteria, and there are indeed some instances
where astronomical lines may have been preferentially selected. For example,
at some sites only the indication one way along an alignment has been
included, and not the other - we must check that this decision was not
based on the first line being astronomical and not the second. At other
sites certain inherently likely indications per se, such as lines along
alignments of stones, have been omitted in favour of intersite lines, and
again we need to check why the selection was made.

There are also a few archaeological misinterpretations included
amongst the alleged indications. However these only represent a small
minority of the lines, and simply require identification and removal.

A problem of more fundamental importance is the inclusion at
Level 1 of sightlines to what are considered to be indicated horizon foresights, as distinct from indications purely on the ground. Something like
1/5 of the declinations included in the later (Thom 1967a) analysis are
in fact the declinations of foresights assumed to be indicated by some
simple indicator on the ground. In these cases we are often not told what
the claimed method of indication is; we are often not told what the foresight is; and we are never told how wide the indication is, and how many
other equally prominent and well-indicated horizon foresights might equally
well have been chosen by the investigator.

What is happening here is that two distinct hypotheses are
being muddled. One is that structures on the ground, such as stone alignments, themselves provide an astronomical indication, and that the actual
shape of the horizon they point to is irrelevant. The other is that
natural horizon foresights provide the astronomical sightline, and all the
structure on the ground has to do is to point out - preferably uniquely -
which foresight is to be used. These are hypotheses we ought to test in
separate analyses - although there is nothing to stop us using the same

data base for both. The danger is that we do the following: if there are
no notches in any given indicated stretch of horizon, we conclude that if
the line was significant at all it could only have been of the "indication
on the ground" type; so we put in to our analysis a gaussian hump centred
upon the centre of the indicated horizon. But if we find one or more
notches within the horizon range (or maybe even somewhat outside it), we
might conclude that the sightline, if significant at all, is of the indicated foresight type. The possibility of selective bias enters if our
decision about whether any line was an indicated foresight or not is
influenced not just by whether there is a notch there, but by what we find
the declination of that notch to be. We might survey all the notches
present, and if any turn out to yield an astronomical declination, we conclude that the line was an indicated foresight, and include the declination
of the relevant notch in our analysis. If on the other hand no notch
turns out to be astronomical, we conclude that the line was at best only
a rougher "indication on the ground", and simply include a rougher declination corresponding to that. The analysis of an effectively arbitrary
selection of 38 of the 262 sightlines included by Thom (1967a, table 8.1), those
giving declinations near to the lunar extremes, shows evidence for this
type of decision having been made (Ruggles 1981, section 4.3), and it
seems reasonable to assume that Thom felt quite justified in doing so.
However, the result will inevitably have been to introduce considerable
selective bias into the data. Instead, we need to test separate hypotheses
separately, if we are to prove that astronomical trends are really present
in the site data, and that we are not simply fitting the most convenient
theory at each site.

A final point at Level 1 is more simple: it is that given we
are dealing with a number of different types of indication, the accuracies
of which vary, this should be reflected in different widths of gaussian
hump, not just two fixed widths as used by the Thoms. Possibly a better
procedure would be to spread a fixed probability evenly over the possible
range of declinations for each sightline. For instance, a single stone
slab, especially if it is leaning or badly weathered, might originally
have been meant to point anywhere in an azimuth range some degrees wide.
In that case we should spread the probability evenly over the whole range,
rather than choosing a horizon point somewhere in that range that we assume
to have been indicated; for otherwise our personal interpretation will
enter, and we again risk being influenced by astronomical considerations.

Because of the large number of sites involved in Thom's later
analysis at Level 1, there seems little point in simply revisiting and reassessing the sites considered in that analysis, as has been done at the
higher levels. It would be much more constructive to attempt to formulate
a more explicit site methodology, and to collect site data independently
of Thom. In this manner one can use his overall results in formulating
rigorous selection criteria (that is, in deciding what are the "most
likely" indications at any particular type of site, so as to give oneself
the best possible chance of a positive result), but at the same time trying
to be uninfluenced by his interpretation of any particular site. A project
along these lines forms part of the work described in the second half of
this paper.

METHODOLOGICAL SUGGESTIONS AND NEW WORK
General considerations

What follows is an attempt to identify some of the general
problems involved in analysing the orientations of archaeological sites,
starting from scratch and disregarding (for the moment) the groundwork of
Thom. These general remarks could apply equally well to a variety of types
of site: to ceremonial centres and cities in Mesoamerica (e.g. Aveni
1975) or to grave orientations from the British Earlier Neolithic period
(Ashbee 1970) through to Roman and Saxon times (Faull 1977; Rahtz 1978)
as much as to British megalithic sites.

Orientation hypotheses can be broadly separated into three
categories, with the following general headings covering those factors
which might be hypothesised to determine or affect structure orientations.

> (i) Astronomical considerations. Structures might be aligned
> upon the heavenly bodies, e.g. towards the horizon rising and
> setting points of the sun at a solstice or other significant
> time in the year.
> (ii) Compass directions. Structures might be deliberately or
> preferentially aligned north-south, or in some other direction.
> (iii) Features on the ground. Structure orientations might
> be towards nearby sites of a certain type, natural features
> such as distant mountains, or directions of local significance
> (e.g. whence ancestors came); or else they might depend upon
> the local lie of the land (e.g. tomb entrances downhill).

The testing of any particular causal hypothesis must depend upon a number
of directions being deemed "significant". In cases (i) and (ii) universally significant directions may be defined, in terms of declination and
azimuth respectively, but in case (iii) significant directions under any
particular hypothesis need to be identified for each site individually.
Our first procedural problem, then, is how many directions to consider
significant. If too few, trends of possible importance may be missed. If
too many, then so many fortuitous orientations will be considered significant that they will inevitably overwhelm any genuine trends in the data.
In the astronomical case, for example, the inclusion as significant of all
stars of at least first magnitude (i.e. the 15 or so brightest stars) over
a 500-year timespan covers about 1/3 of the horizon. Fortunately, in
cases (i) and (ii) we can consider "free" hypotheses: that is (to take
case (i)) we can merely hypothesise that some orientations were astronomically determined, without specifying by which heavenly body, and leave it
at that: we then simply watch for any statistically significant cluster-
ings to build up at certain declinations as data from more sites are added.

Any given orientation will, presumably, actually have resulted
from any number of different factors, acting together or vying against one
another in importance: for example, it might have been desired to situate
a ritual site in a position with wide views, within the restriction of the
builders' available territory; a desired solar or lunar rising or setting
alignment behind a distant mountain, for ceremonial reasons, might have
been secondary to the other requirements, and a compromise reached. In
the absence of independent evidence, we can never prove that any particular structure orientation (that of the Newgrange roof-box and passage
(Patrick 1974), for instance, which is aligned upon midwinter sunrise) was
motivated by a single overriding consideration: it might have arisen
through the chance combination of other factors, many of which are inacc-
essible to us. For example, sheer human perversity might have operated,
as is known from some ethnographic cases (e.g. Burl 1981, pp.250-1).
Thus we are of necessity forced to seek a large data sample and to use
statistical inference on the sample as a whole, hoping that the causal
factor being tested will lead to trends distinguishable at high statistical significance levels from the effects of alternative factors.

In practice one tests any particular orientation hypothesis
against the alternative that the orientations in question were randomly

distributed. It is essential, then, to check (as far as is possible) that
alternative causal factors would merely lead to random background noise
amongst the data being tested. There may, of course, be many possible
causal factors which, not now apparent to us, we would never think of
testing; however any reasonable known alternatives must be examined. If
this check is not made, demonstrably invalid conclusions can result. As
an example, consider the group of earthen long barrows on Cranborne Chase
examined by Ashbee (1970, pp.23-4). Three quarters of them face south-
easterly directions, a fact which could undoubtedly show up as significant
under a type (ii) hypothesis. However a type (iii) hypothesis fits the
data just as well: it is that the builders were compelled to lay out the
mounds along the ridges upon which they were erected (Ashbee 1970, p.24).
This causal factor does not produce random azimuths because the sites all
lie on predominantly NW-SE ridges.

A further point to be made at this stage is that structure
orientations randomly distributed in azimuth do not produce a uniform
distribution in declination: declinations close to the colatitude are
more probable, because lines of declination graze along the horizon in
the north and south. Thus in testing a type (i) hypothesis a "compress-
ion factor" (Cooke et al. 1977, p.125) must be employed to cancel this
effect.

Having accumulated orientation data from a large number of
sites, we encounter the problem of its statistical analysis. A major
concern is that we do not know how precisely any given orientation, if
deliberate in terms of the hypothesis being tested (e.g. astronomical),
was set up, or was intended to be set up but for the operation of com-
plicating factors on site. Neither can we expect the intended precision
to be uniform from site to site. As a first approximation it seems
reasonable to test for a fixed intended precision in all the structure
orientations considered, but the actual value of this intended precision
is an unknown which must be allowed for in our testing. A method of doing
this has been put forward by Freeman & Elmore (1979) which basically
involves displaying the result of the statistical test on a graph against
a range of values for the maximum allowed deviation of any given orientation from a significant or "target" value. The point here is that an
astronomical enthusiast testing for astronomical significance might well
present only that value for the maximum allowed deviation which gives the
highest apparent statistical significance for his results, whereas a

sceptic might well present only that value which gives the least apparent significance. In order to be unbiased we need to examine the whole range of possibilities in some formal manner.

If a hypothesis (say an astronomical one) is found to be significant we are then in a position to return to the data in more detail and investigate possible variations in astronomical precision with site type or geographical location, in the hope of correlating evidence on variations in the practice of astronomical orientation with other facets of archaeological evidence about the sites.

A good deal of development is needed in the techniques of rigorous statistical analysis in order to deal with different types of orientation data. Certain problems - such as the fact that different orientations at certain sites equally plausible per se (e.g. both ways along a line of stones), can not be considered statistically independent - mean that the techniques of Freeman & Elmore require considerable development and refinement before they will deal satisfactorily with much new archaeoastronomical data. For the time being, though, the important priority is to concentrate upon techniques of data collection in a sufficiently objective manner and upon seeing whether any results actually merit the development of the relevant statistical techniques.

Confined groups of demonstrably similar sites

Having established the need to seek a large data sample from many sites, we encounter the problem of selection of material. In an ideal case we would be provided with a well-defined group of demonstrably similar sites confined to a given area, but sufficient in number to provide a reasonable data base, and with a design such that one direction at each site (e.g. that of the entrance of a chamber tomb) is obviously of special importance. The hope would be that trends amongst such groups could lead to conclusions at high statistical significance levels. Two promising groups have been investigated recently by Burl, on the basis of existing survey data: the Recumbent Stone Circles (Burl 1980) and the Clava cairns (Burl 1981, section 7.5) of north-east Scotland. Both groups exhibit manifest orientation trends (Burl 1976, fig. 25) which seek explanation. In the case of the Recumbent Stone Circles, Burl (1980) has hypothesised that the azimuths of the recumbent stone centres as viewed from the circle centre fall within a defined range of lunar significance rather than aiming at particular rising and setting points. This is a

theory requiring verification of the field data by site surveys of uniform
quality, and a more rigorous hypothesis-testing approach both in the
collection and reduction of this data.

During the summer of 1981 all 50 or so Recumbent Stone Circles
in sufficiently good condition have been surveyed by the author, partly
in collaboration with Burl. Measurements recorded included not only the
azimuths of the recumbent stone centres as viewed from the site centres,
but also various other orientations of possible significance, such as the
azimuths of the inner edges of the flanking stones as viewed from the site
centre, from the diametrically opposite side of the circle to the recumbent stone, and from points perpendicular to the orientation of the recumbent stone. Such points can only be defined with varying reliability from
site to site. In the case of the site centre, this is because many or
all of the circle stones have often been moved or removed, and inner ring
cairns may or may not exist. Such variations must be explicitly identified
and acknowledged, otherwise there is a danger that an investigator might
make assumptions (e.g. about the position of the centres of sites in poor
condition) which are not obvious in his published account and might have
been influenced by his own favoured hypothesis about the function of the
sites. Survey data was also collected to test related hypotheses of
interest, for example that the top of the recumbent stone preferentially
stood below the skyline as viewed from certain positions, and that sites
were preferentially positioned with a distant view over the recumbent
stone. It is hoped that preliminary results will be available early in
1982.

Many groups of monument - chamber tombs, cairns, henges, and
non-circular stone rings, for example - might be susceptible to similar
study. The methods of data collection are non-destructive and economic-
ally viable in these days of financial restraint. Some indication of the
potential of such work will doubtless be gained from the outcome of the
pioneering work on the Recumbent Stone Circles.

Complete samples of sites in certain areas

There remain large numbers of British megalithic monuments,
such as isolated megalithic rings with outliers, standing stone alignments
and single standing stones, which do not fall into large well-defined
geographical groups. They may represent a variety of motivations and span
a long period in terms of culture change. Many of them are of a design

(e.g. three or four stones, not in line) which does not suggest any unique, overridingly important, site orientation. Some of them may be partially destroyed, further standing stones having vanished without trace. Finally some may not be genuinely prehistoric, such as single standing stones erected in more recent times as trackway markers, boundary stones or cattle rubbing posts. Might there be any value in a large-scale study of such monuments? The partially destroyed and bogus sites would introduce an unknown amount of "random noise" into the data, even if many prehistoric structures had been deliberately oriented. In addition, it would surely be over-optimistic to hope that more than a small proportion of the genuine sites, being of such variety and erected over a timespan of something like 1500 years, could have their orientations principally determined by any particular cause, such as astronomical considerations. In other words, the study of the orientations of complete samples of megalithic sites (other than tombs) in certain areas is a venture which seems largely without hope of reward. Or at least it would seem so, were it not for the early work of Thom. For such an approach is exactly that employed by Thom at what we have called Level 1. At the very least it has to be admitted that in Thom's earliest work, where his selection   methodology was most explicit (Thom 1955), significant evidence seems to be present for solar solstitial and equinoctial orientations and alignments. This in itself - quite apart from the far greater implications of the later Level 1 work (Thom 1967a, ch. 8) and the results Thom obtained at Level 2, whose explaining away presents some problems (see above) - demands that we attempt to move on from studying groups of demonstrably similar sites (which might be termed "Level Zero") to Thom's Level 1.

A central problem in doing so is how to select structures for inclusion in the analysis at each site, when a variety of types of site are being considered. In order to prevent the investigator making his own decisions on site, and thereby possibly admitting unconscious bias, a predefined code of practice must be laid down, which must be flexible enough to cope with each different type of site encountered. It must be just sufficiently selective so as to provide enough data for the analysis, while at the same time not allowing any potentially significant evidence to be submerged and lost amidst a welter of data which is patently irrelevant to the hypothesis. As an example, consider the following selection criterion: "the orientations of all lines joining two standing stones at a site are to be included in the analysis (in both directions)".

At a site consisting only of two menhirs this is a good criterion, as these
are the most obvious ways in which an orientation at the site might be
significant. However now consider a circle of 20 stones: the criterion
would have us include from the site 380 orientations, virtually all without a doubt of no particular significance. The author and his colleagues
J.A. Cooke, R.W. Few and J.G. Morgan (Cooke et al. 1977) have attempted
to lay down a code of practice by classifying likely structures for deliberate astronomical orientation into an order of preference, taking into
account the frequency with which they are mentioned by Thom (1967a, ch. 8),
and then only considering as possibly significant at any particular site
those with the highest classification that exist there. Obviously many
different classification sequences might have been chosen, but our particular choice was made because, in the light of Thom's work, it seemed
most likely to yield positive results. Since their publication in 1977,
these selection criteria have been rightly criticised as being too res-
trictive (e.g. Heggie 1981b, p.141), and in some cases questionable (e.g.
Heggie 1981a, p.S19), and in subsequent work they have been modified to
take account of such criticisms.

There are a number of minor problems, such as procedure at
sites where the investigator finds fallen stones whose original position
is unknown, or partially submerged stones which may or may not be fallen
menhirs. In order to prevent the investigator being left to make (possibly subjective) judgements, the code of practice must cater for these
cases also. One possibility is to attach lower statistical weight to
lines involving fallen or dubious stones, another to give them a lower
classification on the classification scale, and another to assign them a
wider range in azimuth, so as to spread out the probability more than
would be the case with better-defined orientations. A combination of
the second and third options is favoured by the author.

Between 1976 and 1978 over 100 megalithic sites were visited
and surveyed by the author and his colleagues J.G. Morgan, R.W. Few,
J.A. Cooke and S.F. Burch. The areas chosen for study were the islands
of Mull (1976), North and South Uist (1977) and Islay (1978). During 1979
and 1981 the number of sites studied was increased to some 300, and further
areas covered were Kintyre, Knapdale, mid- and north Argyll, the islands
of Jura, Coll, Tiree, Barra, Lewis and Harris, and parts of Orkney and
Shetland. Several areas and sites were visited more than once, where it
was desirable and convenient. Most of this data is at best only partially

reduced, but it is hoped to have most of the results available by the
summer of 1982. Eventual publication of the work will involve an archaeological appraisal of the sites, the reduced survey data as appropriate,
and the results of testing the hypothesis that many structure orientations
were influenced by astronomical considerations. One product of this work
will be a new overall declination histogram, which will represent a
similar approach to that of Thom (1955, fig. 8; 1967a, fig. 8.1), but
with the field data collected independently and hopefully in a manner less
susceptible to the objections raised in the first part of this paper.

FINAL REMARKS

It would seem unduly optimistic to expect any more to emerge
from the Level 1 project just described than some tentative pointers to
future work. Overall, we may find statistically significant evidence for
certain types of astronomical orientation, as did Thom, by considering
together hundreds of diverse megalithic sites. Any conclusions that
follow will, of course, be of a very general nature. Beyond this, we can
hope for the emergence of trends above the random data, which relate to
small groups of similar sites in certain areas; for example, we might find
that the small number of three-stone rows in northern Mull seem to have
certain astronomical orientations in common. This would encourage us to
study the relevant sites more closely, to examine or re-examine sites of
similar form in other areas, and so on. But because of the small size
and uncertain extent of such groups of similar sites; the fact that a
number of sites are of a form only repeated at most in one or two other
instances (such as the unique nature of the main site at Callanish); and
that any particular form of site may have resulted from different periods
of occupation and phases of construction in prehistory, and uncertain
partial destruction since, we can not expect to collect enough reliable
data for any conclusions to be statistically testable on the more specific
level. It seems highly likely, then, that any more specific inferences
from work at Level 1 will be of significant archaeological value only in
the context of much better-established results from studies of large groups
of demonstrably similar sites, "Level Zero", as discussed previously.

I believe that only when we have done sufficient work on demonstrable rougher site orientations, and only when we have started to
relate convincingly the astronomy to the archaeology at those groups of
sites, like the Recumbent Stone Circles and the Clava cairns, where we

have available a good deal of archaeological evidence, will we be properly
prepared to return to Level 1 and draw any meaningful conclusions about
the astronomy of those sites such as stone alignments about which we know
far less archaeologically. Archaeoastronomy could well be a useful tool
for investigating unexcavated sites, but it has to be used with a very
great deal of care. An essential prerequisite to its more widespread use
is an explicit methodology. For one thing, it has to be seen as one
facet of orientation study in general. And for another, it is valueless
if orientations are studied in isolation. The archaeoastronomer must
continually strive to marry together orientation evidence with all the
other strands of archaeological evidence before he draws his conclusions.
After all, this sort of joint approach is surely what archaeoastronomy
<u>should</u> be all about.

REFERENCES

Ashbee, P. (1970). The earthen long barrow in Britain. London: Dent.
Aveni, A.F. (1975). Possible astronomical orientations in ancient Meso-
          america. In Archaeoastronomy in pre-Columbian America, ed.
          A.F. Aveni, pp.163-90. Austin: University of Texas Press.
Burl, H.A.W. (1976). The stone circles of the British Isles. London &
          New Haven: Yale University Press.
Burl, H.A.W. (1980). Science or symbolism: problems of archaeo-astronomy.
          Antiquity, 54, 191-200.
Burl, H.A.W. (1981). "By the light of the cinerary moon": chambered
          tombs and the astronomy of death. In Astronomy and society
          in Britain during the period 4000-1500 BC, eds. C.L.N. Ruggles
          & A.W.R. Whittle, pp.243-74. Oxford: British Archaeological
          Reports, B.A.R. 88.
Cooke, J.A., Few, R.W., Morgan, J.G. & Ruggles, C.L.N. (1977). Indicated
          declinations at the Callanish megalithic sites. Journal for
          the history of astronomy, 8, 113-33.
Faull, M.L. (1977). British survival in Anglo-Saxon Northumbria. In
          Studies in Celtic survival, ed. L.R. Laing, pp.1-55. Oxford:
          British Archaeological Reports, B.A.R. 37.
Freeman, P.R. & Elmore, W. (1979). A test for the significance of
          astronomical alignments. Archaeoastronomy, no. 1, S86-96.
Heggie, D.C. (1981a). Highlights and problems of megalithic astronomy.
          Archaeoastronomy, no. 3, S17-37.
Heggie, D.C. (1981b). Megalithic science. London: Thames and Hudson.
Morrison, L.V. (1980). On the analysis of megalithic lunar sightlines
          in Scotland. Archaeoastronomy, no. 2, S65-77.
Patrick, J.D. (1974). Midwinter sunrise at Newgrange. Nature, 249,
          517-9.
Rahtz, P. (1978). Grave orientation. Archaeological Journal, 135, 1-14.
Ruggles, C.L.N. (1981). A critical examination of the megalithic lunar
          observatories. In Astronomy and society in Britain during
          the period 4000-1500 BC, eds. C.L.N. Ruggles & A.W.R. Whittle,
          pp.153-209. Oxford: British Archaeological Reports, B.A.R. 88.

Ruggles, C.L.N. (1982). A reassessment of the high precision megalithic
          lunar sightlines, Part one: Backsights, indicators and the
          archaeological status of the sightlines. Archaeoastronomy,
          no. 4, in press.
Ruggles, C.L.N. (1983). A reassessment of the high precision megalithic
          lunar sightlines, Part two: Foresights and the problem of
          selection. Archaeoastronomy, no. 5, in preparation.
Thom, A. (1955). A statistical examination of the megalithic sites in
          Britain. Journal of the Royal Statistical Society, A118,
          275-95.
Thom, A. (1967a). Megalithic sites in Britain. Oxford: Oxford University
          Press.
Thom, A. (1967b). Contribution to "Hoyle on Stonehenge: some comments".
          Antiquity, 41, 91-8.
Thom, A. (1971). Megalithic lunar observatories. Oxford: Oxford Univer-
          sity Press.
Thom, A. & Thom, A.S. (1978a). Megalithic remains in Britain and Brittany.
          Oxford: Oxford University Press.
Thom, A. & Thom, A.S. (1978b). A reconsideration of the lunar sites in
          Britain. Journal for the history of astronomy, 9, 170-9.
Thom, A. & Thom, A.S. (1980). A new study of all megalithic lunar lines.
          Archaeoastronomy, no. 2, S78-89.
Thom, A.S. (1981). Megalithic lunar observatories: an assessment of 42
          lunar alignments. In Astronomy and society in Britain during
          the period 4000-1500 BC, eds. C.L.N. Ruggles & A.W.R. Whittle,
          pp.13-61. Oxford: British Archaeological Reports, B.A.R. 88.

ASPECTS OF THE ARCHAEOASTRONOMY OF STONEHENGE

R.J.C.Atkinson
Department of Archaeology, University College, Cardiff, Wales

Abstract. The building history of Stonehenge is outlined, and
the principal astronomical hypotheses relating to each period
of construction are critically examined. The conclusion is
drawn that only the alignment of the Avenue on the summer
solstice sunrise, established in period II at the end of the
third millennium BC, can be accepted with confidence. All
other interpretations are open to doubt or to alternative
explanations. A note is appended on the possible equinoctial
alignment of the West Kennet Long Barrow, dated to the mid-
fourth millennium BC.

Stonehenge has a history of construction and use extending
over more than two thousand years, from about 3100 BC to about 1000 BC or
later, on the evidence of radiocarbon dates. As an aid to the understanding of the sequence of building, and of the possible astronomical alignments incorporated in each phase of construction, the principal components
of each period are summarised below, with the relevant dates corrected
according to R.M.Clark (1975), and are illustrated in figs. 1 and 2. A
more detailed description can be found in Stonehenge (Atkinson 1979).

The earliest structures at Stonehenge are three pits found
during the extension of the car park. Their excavators (Vatcher & Vatcher
1973) believed them to be postholes for massive tree-trunks, although
their form and cross-section differ from those of other excavated post-
holes at Stonehenge. The late C.A. ('Peter') Newham suggested that they
could have served as astronomical foresights (Newham 1972); but this must
be discounted because pine charcoal from two of them has yielded radiocarbon dates of 7180 bc ± 180 and 6140 bc ± 140 (L.Vatcher, personal communication). These holes are thus apparently of Mesolithic date, and have
no significance for the later Neolithic and Bronze Age structures of
Stonehenge. They will not be referred to further.

Period I (c. 3100 BC) comprises the circular ditch with
internal bank; the ring of 56 Aubrey Holes; the array of postholes on the

entrance causeway, with inside them a pair of stoneholes (D,E) astride the axis of symmetry; the four postholes (A) outside the entrance, which are symmetrical to the same axis; and probably the newly-discovered stonehole (97) close to the Heel Stone (96) (Pitts 1981 a,b).

Fig. 1. Plan of periods I and II.

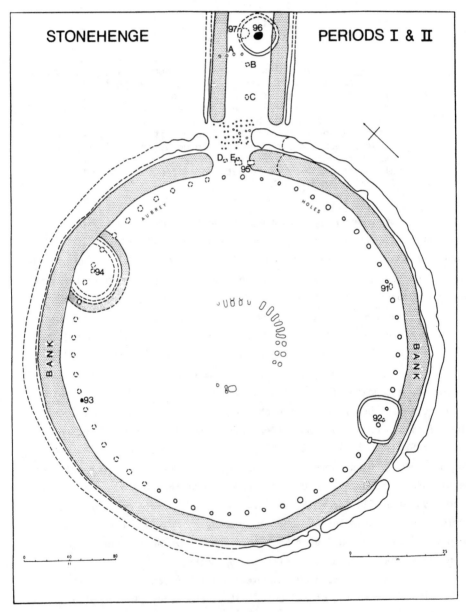

In period II ($\underline{c}$. 2150 BC) the erection was begun in the centre of the enclosure of period I of two concentric circles of bluestones, originating ultimately in South Wales at a distance of 215 km from Stonehenge but probably fetched proximally from a site, as yet undiscovered, in west Wiltshire. The stoneholes so far excavated show that this double circle had an entrance on the north-east side, flanked by additional inlying stones, and a single large stone, possibly the present Altar Stone, diametrically opposite the entrance. At least one quarter of the circumference of both rings had not been completed when the erected stones were dismantled at the beginning of period IIIa.

At the same time the axis of symmetry of the monument was moved about 3^0 clockwise by throwing back the already eroded bank into the silted ditch for a distance of about 8 m on the east side of the entrance. On this new axis the nearer portion of the Avenue was built, running downhill in a straight line to near the bottom of a dry valley for a distance of 510 m. The pair of stones which formerly stood in holes D and E were probably then moved to the holes B and C on the mid-line of the Avenue. It is likely, but not certain, that the four Station Stones (91-94) were erected during this period, together with the Heel Stone (96) and its surrounding ditch, after the dismantling of stone 97.

Period IIIa ($\underline{c}$. 2000 BC) comprises the outer circle of thirty sarsen stones, originally capped by a continuous ring of sarsen lintels, enclosing a horseshoe of five sarsen trilithons, together with a pair of close-set sarsen pillars in the entrance of the earthwork of which the fallen Slaughter Stone (95) survives. The sarsens still standing are shown in solid black on the plan of period IIIb (fig. 2). Their origin lies on the Marlborough Downs some 32 km to the north of Stonehenge.

In period IIIb (date unknown) the evidence suggests that 22 bluestones, probably the largest, were selected and dressed to shape and then erected in an elliptical setting within the sarsen horseshoe. This must have included at least two bluestone trilithons, the lintels of which survive, re-used as pillars in the circle of period IIIc, together with three uprights, now in the horseshoe of the same period, whose tops bear traces of defaced tenons. The plan of this setting is restored in fig. 2, with the excavated stoneholes shown in solid line.

Probably at the end of this period, about 1550 BC, the two rings of Y and Z Holes were dug as sockets for the remaining bluestones;

but this project was never completed and no stones were erected, the holes
being left open to silt up over the succeeding centuries.

      Period IIIc (c. 1550 BC) comprises the circle and horseshoe of
bluestones whose remains survive today. Probably at this stage the nine-
teen components of the horseshoe were reshaped to two alternating forms,
a square-sectioned pillar and a tapering obelisk. The Altar Stone stood
as a tall slab on the axis and has since fallen.

      In period IV (c. 1100 BC) the Avenue was extended from its
original termination for about 2.6 km to the bank of the River Avon. This
implies the continued use of Stonehenge as late as 1000 BC and perhaps

      Fig. 2. Plan of periods IIIb and IIIc.

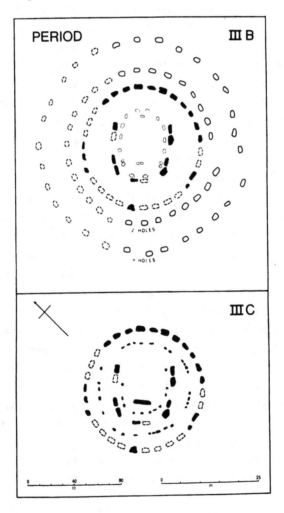

later still. It is significant that air-photographs show no traces of
cultivation of the prehistoric or Romano-British periods in the immediate
environs of Stonehenge, though they are widespread elsewhere in the area.

The axis of symmetry of Stonehenge I, with an azimuth of 46°
33' and an apparent horizon altitude of 36', has no obvious astronomical
significance, nor does that of the estimated centre of stone 97, at 48° 21'
with the same horizon altitude. The ring of 56 Aubrey Holes has been interpreted (Hawkins 1964) as a counter for a 56-year eclipse cycle; but this
has been shown to go out of phase after a few cycles (Colton & Martin 1967,
1969). Hoyle (1977) suggests that this ring could have served as a model
of the ecliptic, with stone markers moved round the circle according to
prescribed rules to represent the longitudes of the sun, moon and lunar
nodes, periodic corrections being made by means of observations on defined
alignments some time before and after the critical events so that the date
of the latter could be more accurately fixed by interpolation. It now
seems likely, however, that the alignments in question belong to period II
and are not part of the same design as the Aubrey Holes.

An even more complex interpretation by Beach (1977) uses the
Aubrey Holes to predict the amplitude of tides in the English Channel from
an assumed knowledge of the occurrence of the lunar apogee; but this must
imply an understanding of celestial mechanics which it is extremely difficult to envisage in a prehistoric context, on the evidence of the Aubrey
Holes alone. The prime objection to all these interpretations is that the
evidence of excavation shows that the Aubrey Holes were deliberately
filled up very soon after they were dug. Moreover there are other sites
in Britain with similar rings of pits containing similar cremated burials
accompanied by similar neolithic grave-goods for which no astronomical
explanation has ever been put forward.

The array of postholes on the entrance causeway has been
interpreted by Newham (1966) as markers for the most northerly rising of
the moon over a series of rotations of the lunar nodes. This is possible;
but clearly the observations thus marked were not made systematically with
respect to time. Alternatively it has been suggested (Thom et al. 1974)
that this array served as an extrapolating sector for determining with
precision the moment of the 'standstill', analogous with fan-shaped rows
of stones elsewhere in Britain and Brittany. It will be seen, however,
that the holes are a poor fit to the theoretical grid superimposed upon

them, even though it is not difficult to position posts of this size to an accuracy of a few centimetres. There may well be other, non-astronomical, purposes for these posts, such as a series of barriers across the entrance causeway.

We owe the earliest recognition of the astronomical alignment of Stonehenge to the antiquary William Stukeley (1740, 35), who said of the Avenue that "it answers ... to the principal line of the whole work, the north-east, where abouts the sun rises, when the days are longest". This was re-affirmed by John Smith in 1771; but it was not until 1901 that the mean azimuth of the Avenue was measured with accuracy. It was then estimated at 49° 35' 51", though a slightly different value was adopted for reasons which can now be seen to have been perverse (Lockyer & Penrose 1902). A more recent survey yielded an estimate of 49° 54' 40", which is close to the theoretical azimuth of the first gleam of sunrise, with 2' of the upper limb visible, if it is assumed that trees 10 m high were growing on the skyline (Atkinson 1978).

This is the one alignment in the structure of Stonehenge which is long enough to be determined with an accuracy sufficient to justify the assumption that it was adopted deliberately. The axis of the contemporary unfinished double circle of bluestones can be estimated only within wide limits of error.

If the Heel Stone and the four Station Stones were erected in period II, the former could not have served as a direct foresight for the summer solstice sunrise, however defined. The latter do appear to have occupied the angles of an approximate rectangle, of which the short sides are aligned roughly on the summer solstice sunrise and the winter solstice sunset and the long sides on the most southerly rising and the most northerly setting of the moon (Atkinson 1978). There can be no certainty about this, however, because the only Station Stone now upright (93) is a stump which may be a replacement of an earlier marker, since its sides are partially dressed to shape in the same manner as the sarsen stones of period IIIa; the other surviving stone (91) is not dressed, but has fallen; and the remaining two are represented only by their stoneholes, which are much larger than the stones set in them.

It has been suggested by Dibble (1976) that the rectangle defined by the Stations was chosen to form a pair of 5, 12, 13 Pythagorean triangles with a common hypotenuse. This is possible within the limits of

uncertainty of the former positions of the four Stations; but it does not
seem possible to satisfy both this condition and the condition that the
sides of the rectangle should align upon the relevant risings and settings
within the date-range of period II.

The axis of symmetry of the sarsen structure of period IIIa
cannot now be recovered with any accuracy, because it was defined by the
centres of gaps between adjacent stones (nos. 15-16, 55-56, 30-1 and E-95),
and none of these pairs is now intact. It may be assumed, however, that it
was intended to conform approximately to the solstitial direction established in period II.

The Thoms' survey of 1973 (Thom et al. 1974) indicated an
azimuth of 49° 57' ± 3' for the long axis of the ellipse which is a good
fit to the inner faces of the outer four trilithons, account having been
taken (personal communication) of the inwards inclination of the second
one (53-54) by 50 mm at the measured level. My own estimate (Atkinson 1978)
is 100 mm, which might increase the calculated azimuth by about 10'. Even
so it is clear that the axis of symmetry of the trilithons, each upright
of which weighs between 40 and 50 tonnes, is astonishingly close to the
best estimate of the axis of the Avenue.

Hawkins (1965) has claimed that the extreme risings and settings of the sun and moon are indicated by alignments of the gaps in the
trilithons with gaps in the outer circle of sarsens. The former are about
0.3 m wide and the latter, at a distance of about 7.5 m, are around 1.0 m
wide, so that even when the radial thickness of the stones is taken into
account the resulting 'windows' cover an appreciable arc of the horizon.
Moreover the sides of the uprights are irregular in profile, so that a gap
at an observer's eye height is not always symmetrically above the gap in
plan at ground level. It is very questionable whether these suggested
alignments are more than an accidental consequence of the lay-out of the
sarsen structure.

Brinckerhoff (1976) has proposed that holes in the upper
surface of the lintels over the entrance of the sarsen circle could have
held sticks or wands marking significant astronomical events for an
observer standing on the lintels on the opposite side, but only if stone
55, the anti-clockwise component of the central trilithon, had already
fallen. This cannot, therefore, be part of the design of Stonehenge IIIa;
and in any case the holes in question are almost certainly the products of

natural weathering. Similar, and indeed deeper, holes occur in other lintels, which can have no possible astronomical significance (Atkinson 1977).

The excavated remains of the elliptical setting of bluestones
of period IIIb, and of the Y and Z Holes, are quite insufficient in number
and too large in size to allow any precise estimate of the azimuth of the
designed axis of symmetry. We can only assume that this was meant to
accord with the axis of the pre-existing sarsen structure, which itself
probably reproduces, within the limits of constructional error, that of
period II. The same is true of the circle and horseshoe of bluestones of
period IIIc.

Thom and his collaborators have suggested that Stonehenge was
the central point of a solar and lunar observatory with foresights up to
14 km away, though without ascription to any specified period (Thom et al.
1974, 1975). Of the six foresights discussed, two are certainly and two
others probably of much later date than Stonehenge itself.

Peter's Mound has been shown by excavation to be the remains
of a military rubbish-dump of the 1914-18 war (L.Vatcher, personal communication). The bank at Gibbet Knoll has far too sharp a profile for it
to be of prehistoric date, and it is in fact the western side of a square
earthwork which was clearly visible as late as 1946, since when the other
three sides have been destroyed by cultivation. It can confidently be
identified as a Civil War gun-battery of the 17th century AD, commanding
the old road from Devizes to Salisbury.

The mound at Hanging Langford Camp is part of a complex of
earthworks known by excavation to date from the late pre-Roman Iron Age,
and there is no reason to assume an earlier date on the evidence of the
alignment alone. The low mound of irregular shape, surrounded by a silted
ditch, in the centre of Figsbury Rings is probably a tree-mound made no
more than 200 years ago. An air-photograph of 1924 (Crawford & Keiller
1928, pl. IX) shows a mature elm tree growing on it, and suggests that the
mound is superimposed on lines of cultivation which can hardly ante-date
the 19th century AD.

The conclusion to be drawn from this review of the major
astronomical interpretations of Stonehenge is that there is only one
built-in alignment which can be accepted with confidence, namely the axis
of the first straight stretch of the Avenue directed upon the first-gleam
solstitial sunrise at the end of the third millennium BC. It is this

alignment, the first to be recognised, together with the unique character of Stonehenge, which has generated the other theories mentioned above, all of which are open to doubt on various grounds. We should remember the wise words of Jacquetta Hawkes (1967): "Every age has the Stonehenge it deserves — or desires". We live in an age of science and technology, and it is difficult to escape from the strait-jacket of our own universe of discourse, let alone to see into the minds and to assess the motives and intentions of our remote prehistoric ancestors on the basis of evidence which is both dumb and ambiguous.

A NOTE ON THE WEST KENNET LONG BARROW

This, the largest neolithic chambered tomb in England and Wales, is one of the complex of monuments near Avebury in Wiltshire. The mean axis of the passage of the tomb has an azimuth of 89º 40' (Piggott 1962, fig. 4). The apparent altitude of the horizon, as measured from the 1:25000 map, is 10.6'. This yields a declination of -17' if sunrise is defined by the first gleam of 2' of the upper limb, and of +10' if the lower limb is tangent to the horizon. Given the minor uncertainties in the estimation of the passage axis, of the horizon altitude including the possible growth of trees, and of the time of sunrise in relation to the time of the equinox proper, a case perhaps exists for supposing that this tomb was deliberately aligned on the equinoctial sunrise, though direct observation is no longer possible because the original blocking-stone has been re-erected and the distant horizon is obscured by trees on a nearer ridge. The date of construction is not directly known; but by analogy with other dated sites it is probably around 3500 BC. The West Kennet Long Barrow could thus be amongst the earliest monuments in Britain for which an astronomical alignment is putatively attested.

References

Atkinson, R.J.C. (1976). The Stonehenge Stations. J.Hist.Astron., 7, 142-4.
    (1977). Interpreting Stonehenge. Nature, 265, 11.
    (1978). Some new measurements on Stonehenge. Nature, 275, 50-2.
    (1979). Stonehenge (2nd ed. reprinted with additions). Harmondsworth: Penguin Books.
Beach, A.D. (1977). Stonehenge I and lunar dynamics. Nature, 265, 17-21.
Brinckerhoff, R.F. (1976). Astronomically orientated markings on Stonehenge. Nature, 263, 465-9.
Clark, R.M. (1975). A calibration curve for radiocarbon dates. Antiquity, 49, 251-66.
Colton, R. & Martin, R.L. (1967). Eclipse cycles and eclipses at Stonehenge. Nature, 213, 476-8.

Colton, R. & Martin, R.L. (1969). Eclipse prediction at Stonehenge.
        Nature, 221, 1011-12.
Crawford, O.G.S. & Keiller, A. (1928). Wessex from the air. Oxford: Clar-
        endon Press.
Dibble, W.E. (1976). A possible Pythagorean triangle at Stonehenge.
        J.Hist.Astron., 7, 141-2.
Hawkes, J. (1967). God in the machine. Antiquity, 41, 174-80.
Hawkins, G.S. (1964). Stonehenge: a neolithic computer. Nature, 202, 1258-61.
        (1965). Stonehenge decoded. London: Souvenir Press.
Hoyle, F. (1977). On Stonehenge. London: Heinemann.
Lockyer, N. & Penrose, F.C. (1902). An attempt to ascertain the date of
        the original construction of Stonehenge from its orientation.
        Proc. Roy. Soc. Lond., 69, 137-47.
Newham, C.A. (1966). Stonehenge: a neolithic observatory. Nature, 211,
        456-8.
        (1972). The astronomical significance of Stonehenge. Leeds:
        John Blackburn.
Piggott, S. (1962). The West Kennet Long Barrow. London: H.M.S.O.
Pitts, M.W. (1981 a). Stones, pits and Stonehenge. Nature, 290, 46-7.
        (1981 b). The discovery of a new stone at Stonehenge.
        Archaeoastronomy, 4, no. 2, 17-21.
Smith, J. (1771). Choir Gaur. Salisbury: Easton.
Stukeley, W. (1740). Stonehenge: a temple restor'd to the British Druids.
        London: Innys and Manby.
Thom, A. et al. (1974). Stonehenge. J.Hist.Astron., 5, 71-90.
        (1975). Stonehenge as a possible lunar observatory. J.Hist.
        Astron., 6, 19-30.
Vatcher, F. & Vatcher, L. (1973). Excavation of three postholes in the
        Stonehenge car park. Wiltshire Arch. Mag., 68, 57-63.

IMPLICATIONS FOR ARCHAEOLOGY

Euan W. MacKie
The Hunterian Museum, The University, Glasgow, G12.

Abstract. Some underlying aspects of the controversies concerning 'megalithic astronomy' in Britain are considered, including possible irrational reasons for supporting and opposing new ideas, the nature of scientific thinking in archaeology, the use of statistics and the value of the evidence of single sites. The use of ethnoastronomical studies, and the methodology of identifying alignments, are each briefly discussed as are the results of some practical tests of the astronomical hypothesis. These last suggest that the evidence for accurate prehistoric observation is weak in Wiltshire but strong in Argyll.

1 INTRODUCTION

For at least fifteen years a considerable controversy has flared intermittently in the field of Neolithic studies in Britain - one which concerns the fairly sophisticated astronomical functions and geometrical qualities attributed to stone circle and standing stone sites all over the country by Alexander Thom. Quite markedly differing positions have been taken up by archaeologists and by other scientists interested in the subject, and the debate shows no sign of dying down; on the contrary phrases like 'semi-lunacy' and 'deluded men' continue to be hurled by defenders of the established order (Daniel 1981, 87).

Anyone with a professional interest in this problem has three broad choices when faced with such a potentially embarrassing situation. Firstly he or she can avoid controversy by silence, or secondly can take up an entrenched and partisan position and keep firing away until the end, whatever happens or, thirdly and if feeling both brave and constructive, he or she can try to study *all* the evidence as dispassionately as possible, to reject *a priori* or *ex cathedra* thinking and to come to whatever sensible conclusion ultimately seems indicated. Part of this process of analysis is to have a clear idea in one's own mind as to what the methods of scientific thinking are, how fact and theory are distinguished in archaeology and how reliable explanations are distinguished from

unreliable.   The first part of this paper considers some of these
problems.

1.1 *Topics to be discussed*

The title refers specifically to the archaeology of Neolithic
and Early Bronze Age Britain - approximately the period from 4000-2000 BC
- because that is the field with which the author is most familiar.
However , many of the general points made apply to other fields in which
archaeoastronomical research has been carried out, and a few apply even
to all disciplines which claim to be based on the scientific thought
process.   Since science in its broadest sense claims to have as a primary
aim the distinguishing between facts and falsehoods, and between reliable,
truthful interpretations and misleading or distorted ones, it appears that
we shall be concerned also with these fundamental ethical problems.   Such
are the consequences when an academic discipline is riven by a controversy
provoked from outside its ranks about the very facts upon which some of
its most cherished assumptions are based.

Two main topics will be considered here.   The first concerns
the personal attitude of scholars to both facts and hypotheses;   this is a
difficult problem, and one to which any one addressing himself invites the
riposte about people in glass houses not throwing stones, but it must be
faced.   The second topic concerns some of the problems involved in the
collection of data for archaeoastronomical research.   As a preliminary it
is worth noting that there are three distinct groups of ideas and
associated evidence in British "megalithic astronomy" which are concerned
with (a) the astronomical skills, or lack of them, present in Late
Neolithic Britain, (b) the geometrical skills, or absence of them,
incorporated in the standing stone sites of that period and (c) the
structure of the society of the time as induced from other archaeological
evidence.   When considering the controversies aroused by the work of the
Thoms , and by that of others in the same vein, it is essential to define
which particular part of this overall scenario of problems the evidence
being argued about refers to.

2 *MENTAL ATTITUDES*

2.1 *Supporting and opposing new ideas*

When a discipline like archaeology is faced with a challenge
like that presented by Professor A. and Dr A.S. Thom its practioners are
in something of a dilemma.   The new evidence is not the kind they are

familiar with and the conclusions seem to fly in the face of the detailed
picture of prehistoric Britain painstakingly built up by many scholars
over many years.   It is surely accepted now that mental processes other
than those based on purely objective, rational thought play an important
part in determining one's attitude to a controversial subject.   Since
understanding what irrational motives exist must help these to be
identified and overcome a few possibilities are offered here.   In
essence there are right and wrong reasons for rejecting and accepting a
controversial new explanation.

The right reasons for rejecting a radical new idea or
explanation are of course the ones that everyone believes they follow.
These are that one can (a) identify flaws in the argument, or (b) identify
important facts which the new hypothesis has ignored or failed to explain
properly, or (c) think of a better explanation for all the data, or (d)
can demonstrate that the new evidence on which the new hypothesis is
based is faulty.   Wrong reasons for rejection, usually subconscious,
can be that one (a) dislikes any major departure from the familiar
scenario one has grown up with, or (b) that one cannot understand the new
evidence on which the new hypothesis is based, or (c) one is frightened
of the disapproval of senior colleagues if one breaks ranks to support an
outsider or a minority view, or (d) that one has few or no original ideas
oneself, or has done little original work, so that one's best chance of
fame lies in assaulting an easy and popular target, or even (e) that one
actually feels threatened and made insecure by all forms of novelty.

What of those who support radical new ideas?   Again there are
right and wrong reasons for this.   If one understands all the relevant
evidence and arguments and believes that the new hypothesis explains
everything better than the traditional one, then support is logically
justified.   However, there are numerous wrong reasons for this attitude
including (a) that one is pre-disposed to unorthodox ideas for some reason,
or (b) that one is insufficiently aware of the evidence that supports the
traditional view, or that (c) one is hostile for some reason to the
'establishment' (archaeological in this case) and wishes to brand it as
incompetent or malicious, or (d) that, being incompetent oneself, one
condemns the whole system in which one is unable to succeed and uses the
unorthodox hypothesis as a weapon in this personal psychological battle.

If one or more of these wrong reasons for supporting or
rejecting a radical new hypothesis is in fact influencing someone's

thinking processes one would expect this to be shown in a simple way.
This has been well described by Eysenck in the context of the continuing
controversy over to what degree people's IQs are determined genetically or
environmentally (Eysenck & Kamin 1981, 157).   It depends on understanding
the difference between the principles of scientific debate - already
alluded to - and, for example, those of legal arguments.   In science the
argument should, according to Popper (1963), follow the course of
conjectures and *refutations* with no-one taking 'sides' and with, ideally,
everyone equally anxious to arrive at the nearest approximation to
'truth' that is possible rather than simply to win the argument.   In
such a context criticism is constructive and should be welcomed by the
proponents of the new hypothesis for whom it will provide vital
assistance in improving their ideas.

However if the protagonists are emotionally involved in a
scientific argument for some reason — perhaps for one of those suggested
earlier — then the debate could well degenerate into one following the
adversary procedures of lawyers, for example.   In legal arguments each
side advances all the facts and arguments which seem to favour its case
and seeks to discredit all those used by the opposition.   'When all else
fails the rule seems to be to abuse the other side's attorney' (Eysenck
& Kamin 1981, 157).   It is not difficult to find striking examples of
the adoption of 'adversary procedures' in the arguments about the Thom
hypotheses concerning megalithic astronomy (e.g. Patrick 1981).

Of course it is possible that another wrong-headed reason for
supporting *and* opposing new ideas may be a tendency to see psychological
motives rather than rationality at work in one's colleagues' thinking
processes but not in one's own!   Even so describing some of the numerous
possibilities for erroneous thinking on both sides may help to create a
calmer climate for debate in future.

2.2   The 'irresponsible negative'

One important harmful phenomenon can be identified which shows
itself frequently when radical new ideas are being evaluated and which is
probably to be simply explained by lack of awareness of the processes of
logical thought.   It has been given the name 'the irresponsible negative'
here.   When a new explanation for some phenomenon is offered and
reactions are required from experts in the field, it is much easier to
adopt initially a sceptical rather than a supporting stance.   Obviously

this is because new ideas must survive constructive criticism if they are
to endure, and the discipline which accepted novelty too easily could
hardly be described as having high academic standards. However, the
negative reaction often seems to go beyond what is required for such a
purpose and in particular sometimes to be based on only a partial
understanding of the new material. In strict logic of course a neutral
attitude should be preserved towards a new hypothesis, and towards the one
it is supposed to replace, *unless* one knows at once of valid reasons (for
example faulty evidence) why the new one must be wrong. It is unlikely,
by contrast, that one will know of valid reasons why the new one must be
right since its testing and full evaluation are not likely to have been
completed when it is first announced. Thus as soon as the new hypothesis
appears from an academically reliable source, and particularly if it is
accompanied by new evidence, then the only fair attitude to take is to
suspend judgment, to remain encouragingly neutral until full tests have
been carried out. This is true even though it may be one's duty to
subject new ideas to close questioning to satisfy oneself that its
author's methods are sound and his knowledge adequate.

Thus until one or other of the two rival hypotheses are
satisfactorily proved correct, and provided they both claim to explain
most or all of the available evidence, they are in fact equally balanced
in the scales of scientific judgment. Thus to say positively that one
thinks the new view is wrong requires as much careful study of all the
relevant material as if one was saying that it was right. Yet too often
one hears negative reactions to new views based on little or no proper
analysis, or on extremely vague generalisations, and it is these worthless
attitudes which come here under the heading of 'irresponsible negatives'
and which are the obvious candidates for labelling as one of the emotional
reactions mentioned earlier.

Of course there are irresponsibly positive actions too, in the
form of worthless explanations advanced on the basis of too little
knowledge, but since these almost invariably originate outside the ranks
of science and scholarship, they are irrelevant to this discussion.
Irresponsibly negative attitudes, by contrast, too often originate within
these ranks.

3 THE SCIENTIFIC METHOD

Another benefit which a controversy such as that accompanying the work of the Thoms should confer is that it should make everyone more aware of the basic principles of those spheres of scientific and scholarly activity which attempt to explain or understand phenomena rather than just collect and classify them.   The first and elementary point always to bear in mind is that there are two diametrically opposed ways of explaining new facts — by *deduction* and *induction*.   Deduction involves using a pre-existing theory to explain something novel, whereas induction — far less common — involves thinking up a new general explanation after considering the facts.   Deductive thinking can be so crude as to scarcely qualify as constructive thought at all, as when new facts are automatically explained by old theories in a sort of reflex action. Usually of course it is a perfectly legitmate thought process with facts of a familiar or trivial kind but it can be dangerously obstructive if it is simply used as an excuse to avoid considering new material imaginatively.

Inductive thinking is obviously the better method to adopt when faced with new or unfamiliar data, and when a new hypothesis is to be generated to explain it.   It is harder however because much more originality of mind, and imagination and effort, are required to invent plausible new explanations instead of just using the old ones.   It should be obvious that deductive reasoning alone is an erroneous means of rejecting new hypotheses about new material!

However, whether an explanatory hypothesis is derived inductively or deductively, if it is to be regarded as scientific rather than speculative it must be tested.   In other words it must *predict* the existence of something somewhere which can then be searched for in the place indicated.   If this something is found the hypothesis is verified and regarded as more likely to be true, especially if more than one such prediction is borne out in this way.   If the prediction fails the hypothesis is falsified and must be abandoned or drastically modified. The simple process of prediction and test is the key to the scientific method and is what distinguishes it from myth-making and other forms of irrationality (Beveridge 1961; MacKie 1981b).   It is perhaps not always realised that the scientific method is basic to practical human experience, being surely directly derived from the first technological thinking — based on trial and error — developed by our earliest tool-making

ancestors.   It might be symbolised in everyday terms by the search for
fault in a suddenly extinguished electric light.   The first hypothesis,
easily tested, might be that the bulb had failed;   found to be incorrect
the next considered might be that the plug fuse had blown, and this too is
easily tested , as are several others until the correct explanation is
found.   The vast fields of non-scientific human thought may be likened to
someone announcing the reason for a failed light and then, either because
he has no knowledge of electricity or is not interested, declining to
examine the apparatus to check!

The point of this very basic and incomplete discussion of
scientific principles is that archaeology — like history, geology and
palaeontology — is an *historical* rather than a *natural science;* in other
words it deals with vanished worlds of the past only random and hopelessly
incomplete fragments of which have survived.   It is therefore much harder
to devise tests for hypotheses about historical scientific evidence —
inevitably based on limited and incomplete data which it usually takes
great labour to obtain — than for those of the natural sciences, the
evidence for which is usually abundant and relatively easily obtained.

An additional major difficulty faced by those practising
sciences concerned with prehistoric or non-historical human societies
(those for which no written or oral records are available) is the
impossibility of directly inducing or deducing social institutions from
mute archaeological evidence (MacKie 1977 a, 7-12).   Since there are as yet
no known laws which can link man's technological and economic practices to
his social institutions, our ideas about prehistoric such institutions
have to be derived by *drawing analogies* with known historical societies
with similar economies and material cultures.   Clearly it is much harder,
perhaps impossible , to test hypotheses about intangibles which are based
on analogies as opposed to those based on direct deduction or induction
from tangible data.

For these various reasons the testing of hypotheses in the
true scientific manner is still not commonly practised in the historical
science of archaeology.   There can be a sharp contrast with the
true natural sciences;   what often happens in archaeology is that a
plausible hypothesis is generated to explain some evidence and *that is the
end of the process*.   It becomes accepted and incorporated in textbooks
and university courses without having been systematically tested.
Paradoxically it then becomes harder to dislodge — probably because,

not having passed through the tempering fire of prediction and test and
thus gained some scientific credibility — it depends much more on the
reputation and personality of the archaeologist who generated it.

All these points help us to understand how the controversy
about the Thom hypotheses has waxed so vigorous. Apart from the question
of statistical significance (section 3.1) a large part of the problem may
well derive from a lack of experience among archaeologists of the true
scientific method as described.

3.1 Statistics and archaeology

However it would be unfair to imply that the incomprehension
is all on the archaeological side. This is far from the case and there
is one particular research technique which can if misunderstood, be
improperly used with reference to archaeological problems, although it
can of course be immensely rewarding and useful. This is of course the
analysis by statistical methods of large quantities of information, a
methodology which is essential in the field of 'megalithic astronomy'
as Heggie has explained (1981). There is no intention here to cast a
doubt on the utility of statistics — that would indeed be a 'retreat into
nescience' (Atkinson 1975) — but merely to try and place the use of
statistics into its correct philosophical framework.

It is commonly asserted by those with mathematical training
who are interested in the problem of 'megalithic astronomy' that sound
statistical analysis of the data is the only way to come to a reliable
conclusion about hypotheses which were themselves first derived through
such analyses. Heggie puts it very clearly (1981, 21). The hypothesis
that many standing stone sites were set up deliberately to form alignments
pointing at astronomically significant points in the sky is a fairly old
one and was at first based on observable facts; many of the monuments do
clearly point to such positions. However an hypothesis which merely fits
the facts is not good enough as we have seen and needs to be tested.

There are in fact two ways of undertaking the vital tests
necessary. On the one hand a practical one could be devised; for
example a site claimed to be an accurate astronomical observatory could
be analysed and, as a result, it could be induced that a physical feature
must exist if the apparatus was to work properly, and this feature could
then be sought by excavation. This was done at Kintraw (MacKie 1974;
1977a, 84; 1981a, 115 & 139) and is being done at Brainport Bay (MacKie

1981a, 128). On the other hand the hypothesis could be analysed by
statistical methods without cutting into the ground.   In this case it
would have to be shown that there are more such alignments than would be
expected to occur by chance, and that is what Thom (1955) tried to do.

The same ideas can be applied to Thom's concept of the
prehistoric unit of length, the 'megalithic yard'; to accept it says
Heggie (1981 , 39) one must find out how well a 'quantum hypothesis'
(the idea that a certain unit of length exists) fits a random set of data
and then see whether the same unit fits the set of diameters of stone
circles better.   If it does the prehistoric 'yard' is acceptable.   The
alternative practical scientific test of this hypothesis is to look for
measuring rods of the right length on archaeological sites, and for
historical evidence of the use of the same or a similar unit elsewhere
(MacKie 1977.a, 53).

From the first, essentially statistical, approach another
idea derives, which is that single sites are unreliable in deciding whether
the Thom hypotheses about megalithic astronomy or geometry are valid.
Obviously one site may appear to be an accurate observatory, or a neatly
set out geometrical figure, quite by chance and it seems reasonable to
insist that a large number should show such features before the
hypothesis is believable.   This point must be briefly considered also.

A fundamental point is the claim that statistics provides the
ultimate test of the validity of these theories after the raw data has
been subjected to a careful re-examination as is being done by Clive
Ruggles (1981).   This is surely wrong because of the simple fact that
mathematics alone is not *automatically* a reliable guide to the real world
but must constantly be checked against that world.   Such checks are the
ultimate tests of the reliability of mathematics, not the other way round.
For example Eysenck provides an interesting illustration of statistically
impeccable nonsense by citing the clear and highly significant correlation
over several years between the number of registered prostitutes in
Yokohama and of pig iron production in Pittsburg in the U.S.A. (Eysenck
1981).   Thus , according to the statistical criteria advanced for
judging the reality of megalithic astronomy, we should be quite satisfied
that the activities of one group of ladies in a Japanese city did indeed
affect the Pennsylvania metallurgical industry in some strange way!
However an elementary knowledge of the real world surely warns us to be
sceptical of this particular statistical 'proof'.

What this surely teaches us is that the demonstration of
high statistical significance is not enough by itself, either to prove or
disprove an hypothesis;   one should not assume that smoothly accomplished
and accurate calculations -- which are performed on homogeneous numbers
derived from inevitably more heterogeneous concrete objects -- *necessarily*
mean that the objects themselves have been manipulated, analysed and
understood in exactly the same way (Moroney 1956, 2-3).   The research is
incomplete and unsatisfactory unless the *practical* scientific method of
testing already referred to is applied to hypotheses based on statistics.
In other words predictions about the existence of physical features must
be derivable from these hypotheses and the existence of these can then be
tested in the standard way -- by fieldwork and excavation or by examining
artefacts.   Moreover such practical tests of hypotheses, if they can be
carried out, should be more reliable and satisfactory than purely
statistical tests.   Any one who doubts this must assume that Pittsburg
pig iron workers have a secret and incredibly fast system of travelling
to Yokohama in their time off!

Perhaps it should be emphasised again that the discussion
above should in no sense be construed as an attack on statistical methods
which are, of course, indispensable to scientific disciplines.   Even
statistically significant correlations which are nonsensical are
presumably to be expected occasionally by the laws of probability but,
by the same laws, most such correlations should reflect a real
relationship, provided that they are based on a reliable and adequate
sample of the facts.   What is being stressed here is that, since the
'facts' are often dubious or ambiguous in archaeology, statistical
'proofs' or 'disproofs' in this discipline are not really sufficient.
Where possible independent physical evidence should be sought which bears
on the problem concerned, preferably in such a way that a prediction made
by the statistically-based hypothesis is tested.   If this is not possible
then all that can be done is to subject the original data to careful
study and to repeat the statistical analyses.

3.2   *The value of single sites*

How does this affect the question of the validity of single
sites?   Evidently in two ways.   In the first place it should depend on
whether the site concerned has been explained as an accurate astronomical
observing instrument simply because its physical features, and the distant

horizon indicated, fit this explanation.   Obviously in such cases no firm
and lasting conclusions could be drawn.   If however the site has been
tested in the practical way described, and the hypothesis confirmed by
the discovery by excavation or otherwise of a predicted feature, then
the matter is entirely different and the site becomes very important
indeed.   Single such sites, with hard physical evidence in favour of, or
against, a particular hypothesis render statistical proof or disproof
much less relevant in their cases.   In fact it would surely be
legitimate to reverse the initial judgment entirely if a single site *was*
found at which the astronomical hypothesis (i.e. that it was an accurate
alignment, or observing *instrument,* not just an approximate orientation)
was tested and decisively confirmed in terms of function and date (which
has not quite happened yet).   This evidence would then outweigh in an
overwhelming manner statistical doubts (Heggie 1981), complaints about
misidentification of sites by the Thoms (Moir 1980, 1981), assertions
that prehistoric Britons did not do that kind of thing but were more
interested in cults of the dead and 'sorcery' (Burl 1980, 1981a, 1981b)
as well as appeals to the 'non scientific' nature of other primitive
astronomical activity (Thorpe 1981).   This in a nutshell is why the
author is seeking such a site and suspects he may have one at Minard in
Argyllshire (MacKie 1981a, 128; section 5.4 below).

　　　　　Moreover it may seem unfair to say so but the reverse
discovery — the finding of a predicted physical feature which showed
that a claimed accurate observatory could not have been one — would not
have quite the same overwhelming implications.   The proof that one site
claimed as such was not a primitive astronomical instrument would be a
jolt to the general hypothesis that such existed, certainly, but could not
prove that such did not exist elsewhere.   Clear proof that even one *did*
exist would demonstrate, by contrast, that the disputed skills and
interests were indeed present at one site in the prehistoric period
concerned and therefore probably were elsewhere also.

　　　　　However, even with sites where it has not been possible to
carry out such a practical test, the principle of fairness in drawing
conclusions alluded to earlier (section 2.2) surely means that judgment
must be suspended.   If it is possible to say, concerning sites claimed
as accurate observatories simply because their visible features fit this
hypothesis, that 'the most significant deficiency in both Thom's and
MacKie's work has been their failure to demonstrate that the agreement

of their data' (concerning Kintraw and Ballochroy) 'is not consistent with
accidental occurrence' (Patrick 1981, 217) then it must also be said that
the principle duty of the critics is to demonstrate that the data *is*
consistent with accidental occurence.   If statistical proof is required
before the hypothesis is accepted as possible then statistical disproof,
not just doubt,is needed if it is to be rejected.   If either such 'proof'
is absent then the astronomical hypothesis remains as logically reasonable
(because it clearly does fit the facts) but awaiting scientific testing.
Such a one is in good and numerous company in archaeology!   In most
archaeological situations, as we have seen, such untested but reasonable
hypotheses are accepted quite happily until a better one comes along.
In the case of Kintraw, however, the criticism quoted above is not
strictly valid because there is some independent evidence that the site
was an accurate observatory (MacKie 1974 ; section 5.2 below).

4   IMPACT OF 'ARCHAEOASTRONOMY'

It is worth looking briefly at the situation in Britain
against a wider background by pointing out that 'archaeoastronomical'
studies have made considerable strides in other parts of the world during
the last 15 years, so that the astronomical hypotheses about the British
standing stones , and the social hypotheses about their designers and
builders , are not something that can be argued about any more in
isolation.   Galvanised by the work of the Thoms and Professor Hawkins ,
field investigators have been looking at the remains of buildings of the
pre-Columbian high cultures of Middle and South America, at the more
primitive stone and earth structures of the Indian peoples of various parts
of the U.S.A., at the Bronze Age temples of Egypt, the Neolithic temples
of Malta, at standing stones in East Africa and at a variety of prehistoric
structures in Central and Northern Europe.   Not all these investigations
will necessarily turn out to be fruitful in the long run — just as the
concept developed by V.G. Childe of an archaeological culture based on
recurring assemblages of artefacts could not be applied usefully to every
kind of ancient site — but it is surely clear that a new window on to the
pre-literate past has been opened and one which will, if used responsibly,
often considerably enlarge our understanding of vanished non-literate
peoples.

4.1 Ethnoastronomical studies

Our knowledge of the astronomical skills and practises of modern primitive peoples has also been steadily enlarged and this of course provides us with the essential background to the archaeological research in this field.   Unless we can find modern primitive peoples with an approximately similar level of culture and technology to the prehistoric ones we are interested in, who also have an astronomical expertise comparable to that deduced for the 'megalith builders' of north-western Europe, then the archaeoastronomical hypotheses about the latter are bound to seem much less plausible.   In fact there is an abundance of such material, much of it collected by Dr. Elizabeth Baity nine years ago in a notable paper (1973) and some of it recently reviewed again by Thorpe (1981).

One of the most interesting pieces of research in this field of 'ethno-astronomy' is that by Dr. David Lewis on the Pacific island navigators, outlined in a paper published by the Royal Society in 1974 and recently expanded in a book (Lewis 197-, 1978).   The problem of how the inhabitants of the remoter islands of Polynesia navigated their canoes across thousands of miles of empty oceans to make landfalls on relatively tiny specks of rock has long intrigued scholars and Dr. Lewis has provided some solid answers, first by finding and learning from the few surviving native navigators and, second, by undertaking experimental voyages in similar canoes himself.

However, the point to reflect on here is the general one that these Polynesian navigators evidently emerged in a technologically primitive society without knowledge of metals and, through need, developed a workable navigational system based on detailed knowledge of the revolving stars.   The techniques used were quite different to those suggested for Neolithic Britain and the motivation was doubtless equally different.   Yet the point surely is that, when the environment was favourable (an immense and usually clear sky) and the need was there, a primitive people produced a skilled class of wise men who developed the necessary expertise by studying the sky, learning from experience and passing on their knowledge to their children by rote learning .   Thus a recent review of this work maintained that Lewis has shown decisively that 'familiarity with the stellar heaven played more than a magico-religious part in the life of ancient societies' (Heyerdahl 1981). So may it have done in Neolithic Britain, though here of course the

missing element in the equation is 'motivation' about which at present
we know nothing and can guess little.   The real problem is whether the
British Neolithic astronomer-priests , or whatever they were, progressed
further to undertaking detailed and accurate observations for 'scientific'
or other reasons.

4.2 Methodology

The question of the application of 'archaeoastronomical'
('astro-archaeological' according to Aveni (1981)) methods in field
archaeology must be discussed briefly.   One frequently-heard objection
from archaeologists is that those investigating astronomical alignments
in ancient structures are often without much background archaeological
knowledge and that there is therefore a risk that astronomical
significance may be imposed on ancient structures by people interested
primarily in alignments and hardly at all in the cultural and
technological background of the sites concerned.   The editor of
Antiquity said in a recent editorial that he admired the Thoms' survey
work but not their conclusions (Daniel 1981, 87).

This risk is also admitted , apparently without being
recognised as such, by Aveni (1981, 1) who has defined 'astro-archaeology'
as 'a field methodology, often operating without cultural precepts, for
retrieving astronomical information from the study of alignments
associated with ancient architecture and the landscape.   It is usually
practised by astronomers, engineers and other people from the hard
sciences who are generally more concerned with the natural world than
with the ancient mind'.

There could be two connected dangers in this view -- one
general and one particular.   The general one is that it seems quite
wrong to go and look for alignments in ancient structures unless one
does have some 'cultural precepts' -- in other words, some idea of what
archaeologists think the sites were used for, how old they are and what
the associated material culture is like and so on.   In fact it is not
possible to search for alignments without preconceived cultural ideas
because the very act involves the assumption that a set of artificial
features was deliberately built in ancient times to form a line pointing
at something.

It must be clear that someone ignorant of archaeological
discoveries could be drastically wrong in interpreting such features as

forming a pointer and could read into the site designs which were simply
not intended by the builders.   We were given good examples recently,
including a wrecked Neolithic chambered tomb on North Uist which was
mistaken for a ruined stone circle by Professor Thom (Moir 1980, 38).   It
should be added that such mis-identified sites — of which there will
always be a few — seem to form but a tiny proportion of the total he has
surveyed.

The lesson for sensible archaeoastronomical research is
surely that it is important *not* to conduct it in isolation — without
'cultural precepts' — but in full co-operation with professional
archaeologists, and by them as often as possible.   It is clear that this
will require a considerable adjustment of outlook, not least among many
archaeologists, but it is no use the latter criticising the excesses of
some archaeoastronomers without doing something constructive to help.
Fortunately this constructive attitude is becoming steadily more common,
as can be seen by the fact that several archaeologists in Britain are now
actively investigating the Thom hypotheses.

One specific danger in holding that 'astro-archaeology' is a
field technique divorced from 'cultural precepts' consists in the
possibility of making mistakes in the identification of, and in the
measuring of the azimuths of, the alignments themselves.   This does not
concern the approximate *orientations* seen in many ancient sites — the
azimuths of the axes of which can be plotted only to within a few degrees
anyway, and which are in any case only claimed to have ritual or
ceremonial, and not necessarily any scientific, astronomical significance —
but of structures claimed to be useful observing *instruments,* and to be
effective calendar markers at least.   Although the Thoms have confined
themselves almost entirely to standing stone sites, others have claimed to
see astronomical pointers in a great variety of artificial structures.
These include the long British prehistoric earthworks known as cursuses,
a variety of features at Stonehenge running from the centre of the site to
outlying stones and posts, from inside a few British chambered tombs out
along their entrance passages, out through the doorways of drystone
tower-forts in Sardinia, in standing stones in East Africa, along the
axes of Neolithic temples on Malta and later ones in Dynastic Egypt, and
of course in a great variety of earthworks and masonry buildings in the
New World.

In all these cases, for the structure (the backsight) to have
served as a useful instrument, the artificial line from it (which, it
goes without saying, must be clearly and unambiguously marked) has to
point to another reasonably distant marker on the horizon, either a slope
or a notch or perhaps an artificial mound or standing stone, behind which
the celestial body rises or sets when seen from the backsight.   In the
case of accurate solstice sites such a horizon mark is essential, and
this may not be easily detectable unless the terrain is rugged.

However, in some of the cases alluded to it is not always
clear either (a) that the builders intended to construct an
astronomical pointer at all or (b) exactly which elements in a complex
structure are defining the line.   A good example is provided by the
Maltese Neolithic temple axes (Agius & Ventura 1981);   not only are a
number of them, as the investigators admit, too poorly preserved for the
sets of door jambs supposed to define the axis to be measured properly
but, even when they are, it is not clear how precisely this line can be
defined on the ground.   An accuracy of 1 minute of arc is claimed for
the resulting azimuths for the best sites but this seems rather
improbable when one has seen these megalithic temples.   Neither does any
information seem to be given about the nature of the horizon visible
along the lines, how far away it is or whether there are any bumps or
notches at the indicated points.

In fact for many of these claimed observatories we cannot tell
from the data presented either (1) whether they are really likely to have
been intended as pointers in the first place or, if they are, (2) whether
they are simply structures with a standard orientation in an astronomical
direction, or (3) whether they could have been useful observing instruments.
In the latter cases the indicated alignment ought to point at a
distinctive horizon marker and, unless this is artificial, *the positioning
of the main site would have to have been chosen in relation to this mark*
and to the chosen celestial body rising or setting behind it.   This is a
very important point which is often overlooked.   Scale drawings, with
altitudes and azimuths, of the relevent parts of the horizon are needed.

Assuming that an homogeneous class of ancient structures has
revealed some standardised potential orientation or alignment built into
them, it should be clear that there are only two ways of demonstrating
that these had astronomical significance and may have been observatories.
The first is to find that the alignments consistently cluster round

astronomically significant *declinations*, (not azimuths) and also, if they
were instruments, that the far ends of the lines are marked on the
horizon in some way.   The second way, which must follow from the first,
is to try to identify sites where the astronomical interpretation can
give rise to a prediction which can be tested by fieldwork or excavation.
These conditions have not always been adequately fulfilled in much
archaeoastronomical research to date.

5   TESTS OF THE ASTRONOMICAL HYPOTHESIS

There are now several sites where the astronomical hypothesis
advanced by Thom has made predictions which have been tested by fieldwork
or other methods; to recapitulate, this hypothesis states that many
standing stone sites indicate *alignments* which could have been used as
accurate observing *instruments* in prehistoric times.   The proven
existence of such instruments would demonstrate the desire for precise
astronomical observation in the Late Neolithic and Early Bronze periods
but would not necessarily tell us why these were needed.

5.1   Stonehenge

Stonehenge is a difficult site at which to demonstrate the
existence of such accurate alignments; although there are many standing
stones and other features which could mark them, the horizon is low, flatly
undulating and nearly featureless so that any long, accurate sight-lines
would have been impossible to arrange unless distant artificial mounds
were constructed as foresights.   Several possible such foresights have
been identified by the Thoms and the dates of three of these have been
checked by fieldwork.   In addition one of the shorter alignments claimed
earlier has been independently dated, and it is this which is commented on
first.

The car park post-holes: In 1966 three large pits were
discovered in the chalk about 285 yds north-west of the centre of the
site; clear traces of the massive wooden pine posts which once stood in
them were noted (Vatcher & Vatcher 1973).   Newham pointed out that several
important solar and lunar setting positions would have been marked by tall
posts standing in these holes when viewed from the Heel stone and the four
Station stones (Newham 1972, 23; MacKie 1977a, 125-6).   However recently
fragments of charcoal found in the post-holes have been dated by C-14 to
the 7th millennium b.c. (Burl 1979, 65); the unpublished dates are about

7180 b.c. and 6140 b.c. Thus the car-park posts were evidently set up
at least 3000 years before the earliest features of the main site were
built and should have no connection with Stonehenge. The fact that they
were of pine also supports this early date since pine grew extensively in
the Boreal period while oak predominated in Neolithic times.

 The midsummer sunrise line: Various claims have been made
about the line running north-east from the centre of the site, which also
forms its main axis. Assuming that the Heel stone is too near and too
inaccurately placed to have been a useful summer solstice marker in the
3rd millennium B.C. (MacKie 1977a, 75-7 & 121-3), a more distant
artificial mound on the same bearing may have served the purpose.
Lockyer supposed that an earthwork on Sidbury Hill 8 miles away was part
of a north-eastern alignment but this is of Iron Age date and much later
than Stonehenge. The Thoms suggested that the small 'Peter's Mound',
about 3000 yards away in the same direction, was an accurate summer
solstice marker (Thom & Thom 1978, 149). The problem is that, if it was,
its position means that it could have served the purpose in the 3rd
millennium B.C. only if the centre of the solar disc was observed against
the mark, a somewhat unlikely procedure. In any case the hypothesis has
been disproved; Peter's Mound was excavated in 1977 and was found to be
modern (Atkinson 1981, 209).

 Long lunar lines: That the builders of Stonehenge I and II
were interested in the Moon's movements was suggested by Hawkins (1965,
fig. 11), in relation to the long sides of the Station Stone rectangle,
and by Newham (1972, 15 & 23), as noted, in relation to the post-holes on
the causeway and in the car-park. Thom and Thom (1978, 151 ff. and
table 11.2) have identified five out of eight such possible artificial
foresights which could have been used to track the Moon's movements from
the main site but three of these have been shown, with varying degrees of
probability, to be recent or modern (Atkinson 1981, 209).

 At a distance of 9.2 miles north-west of the site lies the
Gibbet Knoll, in the right position to mark moonset at its greatest
northern declination; however the earthworks are evidently the remains of
a square enclosure, possibly a Civil War gun battery. Mound M at
Hanging Langford Camp is 8.0 miles to the south-west and suitable as a
distant foresight to mark moonset at its least southern declination; it
has not been excavated but at present there is no archaeological reason
to doubt that it is part of the Iron Age and Romano-British system of

earthworks on the site.   A test excavation on this mound should settle
the matter.   *Figsbury Rings* consists of some earthworks with a small
mound in the centre 6.6 miles south-east of Stonehenge, the mound being
suitably situated to act as the foresight for moonrise at its greatest
southern declination.   However the mound seems to be modern and to lie
on top of traces of ridge and furrow cultivation.

Two more possible distant foresights — as yet untested —
could be somewhere on *Chain Hill* 3.7 miles south-west of the main site
(and suitable for moonset at its greatest southern declination), and
could be the tumulus on *Coneybury Hill,* 0.6 miles to the south-east (for
moonrise at its southern declination).

5.2   Kintraw, Argyllshire

This standing stone and probably associated hill platform
together form a classic indicated winter solstice alignment of great
potential accuracy;   the foresight notch on Jura is 28 miles away (Thom
1967, 155; 1971, 37; MacKie, 1974, 118; 1977a, 84).   The way the hill
platform was discovered by excavation — in the spot predicted by the
astronomical hypothesis — has been described several times and the
counter arguments have recently been assessed (MacKie 1981a, 139).
Attempts have been made to discredit the petrofabric evidence for the
artificiality of the hill platform (McCreery 1980) but without much
success (Bibby 1981).   Its artificiality is clear from surface
indications (Stevenson 1982) although there is as yet no independent
evidence for its age.   The circumstantial evidence for its being
associated with the standing stone, and thereby providing striking
verification (through its discovery in the predicted place) of Thom's
hypothesis of accurate solar observatories, still seems very strong.

However, attempts to find a flaw in the evidence from this site
continue.   For example Patrick has tried to maintain for several years
that it is not possible to see the notch on Jura from the hill platform
(MacKie 1976; Patrick 1981), even though a photograph published once
(MacKie 1977b, 106), and a scale drawing of the horizon made with a
theodolite and published several times (MacKie 1974, fig. 4; 1977a, fig.
14b), show clearly that it is.   In Patrick's view this cannot be right
so the photograph '... has not resolved the conflict, and appears to have
been taken from a position north-west of the author's photograph',
which shows the Jura notch partly obscured (Patrick 1981, 213 & fig. 5.2b).

In the same way the theodolite drawing (clearly marked with scales of azimuth and altitude!) becomes only 'a *sketch* drawing *purporting* to illustrate the view of the notch from the platform' (Patrick 1981, 213: my italics). It is regrettable to see Eysenck's concept of adversary procedures, including this 'abuse of the other side's attorney' (section 2.1 above), so clearly illustrated in an archaeological controversy.

5.3   Cultoon, Argyllshire

This stone circle on Islay has provided independent verification of Thom's hypotheses about prehistoric geometry and measuring systems, as has recently been described (MacKie 1981a, 116).   If the geometry of the circle is beyond dispute it has also provided verification of the astronomical hypothesis about accurate solstice sites but, since there are no visible markers on the ground indicating the mountain peak concerned, it seems wiser to forgo that particular claim here.

5.4   Brainport Bay, Argyllshire

This clear and indisputably artificial alignment is still in the process of being investigated but preliminary descriptions have been published (Fane Gladwin 1978; MacKie 1981a, 128).   Several paved platforms, two small standing stones, a rock notch, a pair of 'observation' boulders and a raised back platform are all lined up on two distance mountain peaks 28 miles to the north-east and over which the solstitial sun still rises (MacKie 1981a, pl. 3.5).   The apparatus seems designed for impressive ceremonial but could easily have been used for accurate observations.   Two C-14 dates are now available, one of a.d. 974 $\pm$ 74 (GU 1000) for the contents of a pit broken through one area of paving and one of a.d. 135 $\pm$ 55 (GU 1434) for charcoal on the prehistoric floor level near the front 'foresight stone'.

The site was diagnosed as an astronomical one in 1977 and since then three sets of outlying standing stones have been discovered by Col. P.F. Gladwin, all at or near apparently significant astronomical directions.   The presence of a thick tree cover makes it impossible at present to see two of these groups of stone directly from the site, or to check whether they are so visible.   The third stone lies by itself among oak trees on a rocky ridge about 225 m north-west of the main site and at about the position of midsummer sunset.   Measurements with a theodolite suggested that the stone was too far to the right to mark this event and

direct observation of sunset on June 21st 1981 confirmed this (the stone
is still invisible but its position was indicated by a bright red marker).

As the sun disappeared on that day one of the observers moved
to the right and stopped, well beyond the main site, when the last gleam
appeared to be over the marker.   At this point was a small semicircular
drystone platform with built edges and abutting against a low rock
outcrop; it was almost concealed by turf and no other artificial
structures are nearby.   Further measurements and observations are needed
but it seems possible that Brainport Bay has produced the second
artificial stone platform in the place predicted by the astronomical
hypothesis.   More discoveries at this site are to be expected.

5.5   Comments

The practical tests of the Thom astronomical hypothesis which
have been carried out by archaeological fieldwork are still few in number
but some conclusions can already be drawn.   Stonehenge has so far failed
to provide any confirmation of the long alignment hypotheses proposed
about its features and two of these have been emphatically disproved,
including the crucial accurate summer solstice marker.   The great site
on Salisbury plain looks increasingly like a monument to Neolithic man's
ritual and ceremonial concerns (which probably included the storage of
some astronomical and geometrical information in the design of the site)
rather than to his scientific ambitions, if any.

On the other hand two sites in Argyllshire — a mountainous
region ideal for the invention and construction of long, accurate
astronomical alignments — seem to be providing very strong support for
the Thom hypotheses concerning solstice observatories; Brainport Bay in
particular is in a class by itself here and seems likely to be a site of
national importance.   It seems indisputable that accurate observing
instruments had been invented, in western Scotland at least, presumably
by about 2000 B.C. , though it is still not clear why they were needed.
No practical tests have yet been carried out on Scottish sites claimed
as Moon observatories.

6   REFERENCES

Agius, G. & Ventura, F. (1981).   Investigations into the possible
        astronomical alinements of the Copper Age temples of Malta.
        Archaeoastron. (bull. for Centre of Archaeoastron.), 4, no. 1,
        10-21.

Atkinson, R.J.C. (1975). Megalithic astronomy: a prehistorian's
        comments. J.Hist.Astron., 6, 42-52.
Atkinson, R.J.C. (1981). Comments on the archaeological status of some
        of the sites. In Astronomy & Society in Britain during the
        period 4000-1500 B.C., eds. C.R.N. Ruggles & A.W.R. Whittle,
        pp. 206-09.
Aveni, A. (1981). Archaeoastronomy. In Advances in Archaeological
        Method and Theory, ed. M.B. Schiffer, vol. 4, pp. 1-77.
        New York: Academic Press.
Baity, E. (1973). Archaeoastronomy and ethnoastronomy so far.
        Curr. Anthrop., 14, 389-449.
Beveridge, W.I.B. (1961). The art of scientific investigation.
        London: Mercury Books.
Bibby, J.S. (forthcoming). Kronos.
Burl, H.A.W. (1979). Rings of stone. London: Frances Lincoln.
Burl, H.A.W. (1980). Science or symbolism: problems of archaeoastronomy.
        Antiquity, 54, 191-200.
Burl, H.A.W. (1981a). By the light of the cinerary moon: chambered
        tombs and the astronomy of death. In Astronomy and Society
        in Britain during the period 4000-1500 B.C., eds. C.R.N.
        Ruggles & A.W.R. Whittle, pp. 243-74. Oxford: British
        Archaeological Reports (no. 88).
Burl, H.A.W. (1981b). The recumbent stone circles of Scotland.
        Sci. Amer., 245 (Dec.), 50-65.
Daniel, G.E. (1981). Editorial. Antiquity, 55, 81-9.
Eysenck, H.J. (1981). Personal communication.
Eysenck, H.J. & Kamin, L. (1981). Intelligence: the battle for the
        mind. London & Sydney: Pan Books.
Fane Gladwin, P. (1978). Discoveries at Brainport Bay, Minard, Argyll:
        an interim report. The Kist, 16, 1-5.
Hawkins, G.S. (1965). Stonehenge Decoded. London: Souvenir Press.
Heggie, D.C. (1981). Megalithic Science. London: Thames & Hudson.
Heyerdahl, T. (1981). With stars and waves in the Pacific.
        Archaeoastron. (bull. for Centre of Archaeoastron.), 4,
        no. 1, 32-8.
Lewis, D. (1974). Voyaging stars: aspects of Polynesian and Micronesian
        astronomy. Phil.Trans.Roy.Soc.Lond., 276, 133-48.
Lewis, D. (1978). The Voyaging Stars: Secrets of the Pacific Island
        Navigators. New York: W.W. Norton & Co.
McCreery, T. (1980). The Kintraw stone platform. Kronos 5 (3), 71-9.
MacKie, E.W. (1974). Archaeological tests on supposed astronomical sites
        in Scotland. Phil.Trans.Roy.Soc.Lond., A276, 169-94.
MacKie, E.W. (1976). The Glasgow conference on Ceremonial and Science
        in Prehistoric Britain. Antiquity, 50, 136-8.
MacKie, E.W. (1977a). Science and Society in Prehistoric Britain.
        London: Paul Elek.
MacKie, E.W. (1977b). The megalith builders. Oxford: Phaidon Press.
MacKie, E.W. (1981a). Wise men in Antiquity?, In Astronomy & Society in
        Britain during the period 4000-1500 B.C., eds. C.R.N. Ruggles
        & A.W.R. Whittle, pp. 111-52. Oxford: British
        Archaeological Reports (no. 88).
MacKie, E.W. (1981b). Too tolerant. Nature, 292 (5822), 403.
Moir, G. (1980). Megalithic science and some Scottish site plans:
        part 1. Antiquity, 54, 37-40.
Moir, G. (1981). Some archaeological and astronomical objections to
        scientific astronomy in British prehistory. In Astronomy &

               Society in Britain during the period 4000-1500 B.C., eds.
               C.R.N. Ruggles & A.W.R. Whittle, pp. 221-42. Oxford:
               British Archaeological Reports (no. 88).
Moroney, M.J. (1956). Facts from Figures. Harmondsworth: Penguin Books.
Newham, C.A. (1972). The astronomical significance of Stonehenge. Leeds:
               John Blackburn.
Patrick, J. (1981). A reassessment of the solstitial observatories at
               Kintraw and Ballochroy. In Astronomy & Society in Britain
               during the period 4000-1500 B.C., eds. C.R.N. Ruggles & A.W.R.
               Whittle, pp. 211-20. Oxford: British Archaeological Reports
               (no. 88).
Popper, K. (1963). Conjectures and refutations. London: Routledge &
               Kegan Paul.
Ruggles, C.R.N. (1981). A critical examination of the megalithic lunar
               observatories. In Astronomy and Society in Britain
               during the period 4000-1500 B.C., eds. C.R.N. Ruggles &
               A.W.R. Whittle, pp. 153-210. Oxford: British
               Archaeological Reports (no. 88).
Stevenson, R.B.K. (1982). More on Kintraw. Antiquity, 56 (March),
               forthcoming.
Thom, A. (1955). A statistical examination of the megalithic sites in
               Britain. J.Roy.Stat.Soc., A, 118, 275-91.
Thom, A. (1967). Megalithic sites in Britain. Oxford: Clarendon Press.
Thom, A. (1971). Megalithic lunar observatories. Oxford: Clarendon Press.
Thom, A. & Thom, A.S. (1978). Megalithic remains in Britain and Brittany.
               Oxford: Clarendon Press.
Thorpe, I.J. (1981). Ethnoastronomy: its patterns and archaeological
               implications. In Astronomy and Society in Britain during
               the period 4000-1500 B.C., eds. C.R.N. Ruggles & A.W.R.
               Whittle, pp. 175-89. Oxford: British Archaeological
               Reports (no. 88).
Vatcher, L. & F. (1973). Excavation of three post-holes in the Stonehenge
               car-park. Wilts.Arch. & Nat.Hist. Mag., 68, 57-83.

PI IN THE SKY

H. A. W. Burl
40 St James Road, Edgbaston, Birmingham

Abstract   (A) Flaws in the methodology applied to the examination of
ancient astronomy in the British Isles are discussed. Study of single
sites may lead to distorted conclusions. So may dates based on stellar
calculations.
      (B) It is recommended that groups of related monuments
should form the basis of future research. Preliminary analysis of
chambered tombs, stone circles and rows suggests there had been a
growing interest in astronomy in early prehistory. The Clava Cairn/
recumbent stone circle tradition in Scotland and Ireland is used as a
model to test this hypothesis.
      (C) The accumulating evidence favours the belief that
alignments were for ritual rather than for an intellectual investigation
of the heavens.

(A) INTRODUCTION "The occurrence of orientation in
prehistoric structures has long been noticed. It has not, however,
received from investigators much more than a passing comment, such as,
'the barrow is directed to the eastward', or, 'the entrance to the
chamber faces the north-west'." (Somerville, 1923, 193).

Sixty years later Somerville would be less disappointed with
the attitudes of archaeologists. He might, however, be disconcerted to
find that much of the effort dedicated to archaeo-astronomical research
is misdirected, excellent in its scientific approach but fallacious in
its interpretations.

In classical Greece scientists curious about natural
phenomena enquired, 'why did it happen?' Finding that such a question
was unanswerable they modified it to, 'how did it happen?' Eventually,
frustrated and humbled, they asked only, 'what happened?' Perhaps in
the British Isles archaeo-astronomers should begin with that modest
question. Instead of assuming that any structure, even a single
standing stone, had an astronomical purpose and that an alignment can be

found if sufficient features are examined, it would be wiser to inspect
whole groups of similar monuments, looking for patterns of orientation
within them.

      It is true that orientation used to be a matter that hardly
concerned archaeologists. 'It will be observed .... that the most
prevalent position of the head is to the west and the east', wrote
Mortimer (1905, xxxvii) rather ambiguously of Beaker burials in
Yorkshire without discussing the problem further. Other archaeologists
have commented generally on the directions in which megalithic tombs
faced (Burl, 1981a, 265) but little research has been done to explain
why there should have been chosen orientations. In an otherwise
stimulating and scholarly book, <u>The Archaeology of Death</u> (Kinnes
et al., 1981), much concerned with grave-goods, tomb structure, social
ranking and funerary rites, not one of the contributors thought it
necessary to ask why there were preferred orientations in cists,
cemeteries and burial mounds. Yet it is obvious that orientation was
a fundamental of funerary ritual.

      Some work has been done on Early Bronze Age burial alignments
(Tuckwell, 1975) but one obstacle in Britain is that we lack large
cemeteries such as Tiszapolgar-Basatanya in Hungary (Kalicz, 1970, 53)
to provide a good statistical sample. The scores of graves at
Tiszapolgar were oriented E-W, men lying on their right sides, women on
their left. 'Such consistent differences in skeletal position are not
found in Early Neolithic burials' (Milisauskas, 1978, 173), a statement
that needs modification if, instead of bodies, Neolithic megalithic
tombs are studied. Barlai (1980, 32) suggested that the funerary rites
at Tiszapolgar were governed by the position of the sun at the time of
the burial. If so, and even Merovingian burials 3000 years later hint
at such a correlation (Fichter & Volk, 1980), then this presents yet
another aspect of archaeo-astronomy for us to consider.

      In the absence of cemeteries researchers in Britain must
study either individual tombs or burials, both of which, fortunately,
can be assembled into groups through their architecture or the grave-
goods, particularly pottery, left with them. The old man under a round
barrow on Roundway Down, Wiltshire (Cunnington, 1857), head to the north,
lying on his left side with his beaker by his feet, would be
astronomically mute were it not that Thurnam (1871, 319) noticed the

custom of early beaker users in Wessex regularly to bury men with heads
to the north, women to the south, quite different from the E-W
disposition of corpses in Yorkshire (Tuckwell, 1975).

Even Stukeley knew this. Following his excavation of a
round barrow in the Stonehenge Cursus group he wrote, 'The body lay
north and south, the head to the north, as that Lord Pembroke open'd'
(Stukeley, 1740, 45), a practice observed in many other Beaker burials
in Europe (Harrison, 1980, 39, 50, 61, 68). So widespread was this
N-S cult that it suggests that early beaker users may actually have been
Beaker people.

It is probable that such orientations were effected
astronomically and that the sun and moon were linked with death in the
minds of the prehistoric inhabitants of the British Isles. The solar
roofbox above the blocked entrance to Newgrange reveals this and so
does the indirect glow of mid-winter sunset in Maeshowe's chamber. One
could seek to explain such associations through anthropological
parallels, knowing that the Rites of Passage and the transmutation of
the dead person to the Other-World has been a worldwide preoccupation
of people at all stages of social development (Turner, 1976; Huntingdon
& Metcalf, 1979).

Yet these patterns can be perceived only when entire groups
of tombs or burials are examined. Newgrange's roofbox by itself could
as easily be interpreted as an 'oracle-hole' for the utterances of a
concealed priest/shaman (Lynch, F., 1973, 158) were it not for the
orientations to solar positions and to cardinal points recorded in the
other passage-graves of the Boyne Valley (O'Kelly, Lynch & O'Kelly, 1978).
These 'tombs' also may have contained the equivalents of roofboxes
(ibid, 339).

Only seldom can we be certain that an 'alignment' in any one
site is anything more than the result of chance. The midsummer sunset
line at right-angles to Ballochroy's midwinter-facing row of stones may
have been just such a coincidence. Many of the other three-stone rows
in that region of Scotland, like Ballochroy, contain rough orientations
on midwinter sunset but, unlike Ballochroy's thin slabs, their rounded
boulders do not permit any convincing, secondary alignment at $90°$ to
the first. The fact that in that latitude midwinter and midsummer

sunsets occur at right-angles to each other may therefore have contributed
to the making of a megalithic myth at Ballochroy. Group study suggests
that it was the midwinter sunset alone that was important in these rows.

Dating of individual monuments by astronomical calculation
can be just as misleading. Based on several faulty magnetic,
architectural and solar premisses Stukeley (1740, 65) stated that
Stonehenge was built in 460 BC. Higgins (1829, 158), using precessional
data, preferred 4000 BC. Lockyer (1906, 67) hypothesised that 1680 BC
was more probable. They were all wrong.

At Callanish in the Outer Hebrides Lockyer (1909, 377)
claimed that the avenue had been aligned on the rising of Capella in
1720 BC. Somerville (1913, 88) amended this to 1800 BC, adding that
the east row of the ring showed where the Pleiades rose in 1750 BC.
Thom (1967, 98) proposed a date of 1790 BC for the avenue and Capella,
and substituted Altair in 1760 BC for the east row. The clustering of
so many dates around 1760 BC is impressive but probably meaningless.
Excavations in 1980-1 at Callanish have recovered early Northern
beaker sherds 'immediately adjacent to the stone ring' (Ashmore, 1980,
5) of a period 400 years or more before the putative stellar rows of
stone could have been constructed. Such archaeological evidence
suggests that these 'alignments' are no more than the selection by
modern researchers of stars which happened to be rising in the right
place at the seemingly right time. Somerville rejected some well-
aligned stars because their dates were 'wrong', and Lockyer actually
cited a later date of 1330 BC for the Pleiades. Almost certainly
Callanish was erected well before 1760 BC which is a time coinciding
with the Bedd Branwen phase of the Middle Bronze Age when the
megalithic tradition was in decline (Burgess, 1980, 23).

A comparable misreading of archaeological material seems to
have been made in their fine fieldwork report by the Pontings (G & M,
1981, 66) at Callanish. They accepted C-14 dates for woodland clearance
and agricultural activity a kilometre from the circle as indications of
when the ring was in use. The dates of 1270±65 bc and 1110±95 bc,
roughly 1490 BC in astronomical years, probably relate, however, to
Late Bronze and Early Iron Age communities. Only thirty miles to the SW
at Northton there had been Late Neolithic and Beaker occupation between
about 2100 and 1900 BC (Simpson, 1976), the pottery akin to the

Callanish beaker sherds which were made around 2300-2200 BC (Ashmore,
pers. comm.). In later years the birch and hazel forest had been allowed
to regrow only to be cleared again by late prehistoric groups during the
centuries indicated by the Callanish C-14 dates. By then the stone circle
may have been abandoned for centuries.

The solution to these inter-disciplinary misunderstandings
must lie in co-operation between astronomers, surveyors, archaeologists
and anthropologists. Too much work in the past has been unrealistic
or nihilistic with much adverse criticism of the work of others
without anything constructive being offered in its place. This
introduction, therefore, will close more creatively than it began by
making three points which, it is hoped, are helpful.

1. For the present we should study groups rather than individual
monuments to avoid the risk of examining an unrepresentative structure.

2. Astronomical alignments are frequently associated with death.

3. In the British Isles there appear to have been three major stages of
astronomical activity extending from about 4500 to 1300 BC. These
stages can tentatively be defined:-

>Primitive. 4500-3000 BC. The time of long burial mounds and
>megalithic tombs.
>
>Developed. 3000-1800 BC. The time of early stone circles
>and henges, built for large gatherings of people.
>
>Local. 1800-1300 BC. Smaller, cruder structures such as
>simple rows of stones put up by family groups.

Although these stages can be detected in all parts of the highlands the
Scottish Clava Cairn tradition is particularly helpful because it extends
from around 4000 BC down to the end of the final phase. Examination of
such a tradition provides a useful test of the wider astronomical
hypothesis.

(B) EARLY PREHISTORIC ASTRONOMY

(B).1 Primitive alignments. 4500-3000 BC. With just two
enigmatically-carved bones from Cheddar (Tratman, 1976) hinting that
Palaeolithic bands in this country may have kept tallies of lunations,
and with nothing at all from the Mesolithic period, it is best to assume
that it is only from Neolithic times onwards that astronomical evidence
will be found. Even then, it is the monuments built of durable stone
that offer the first clues. Any earthen mound is today too weathered for

its exact groundplan to be established and all traces of the majority of
timber mortuary enclosures have long since vanished.

Accordingly, we have to accept that the recovery of archaeoastronomy in the British Isles will be confined to the megalithic
highlands in regions of sparser prehistoric population where the climate,
terrain and pastoral economy created conditions culturally different from
those of the chalklands of the south. It is the stone-built tombs of
those scattered hill-people that preserve the beginnings of astronomy in
Britain but they do not represent the picture of the whole country.

The earliest chambered mounds were probably no more than the
simple burial places of separate families but whether long cairn such as
Monamore on Arran, dated to about 3160±110 bc or 3950 BC, or round
passage-grave like Knowth 8, Co. Meath, 2925±150 bc or 3700 BC, the
builders knew that by custom there was a preferred direction for the
entrance. Every region had its own preference (Burl, 1981a, 265; 1981b,
67). Nearly all the primary tombs faced eastwards. In the Cotswold-
Severn group of which Lambourne, 3415±180 bc or 4300 BC, is the earliest
yet known the entrances faced anywhere between NNE and SSE (Burl, 1979,
95), the famous West Kennet barrow looking directly east towards
equinoctial sunrise, but in SW Scotland the Clyde group with similar
architecture preferred a much narrower arc between NNE and ENE.

Although conclusions must wait upon detailed research it is
reasonable to suppose that such azimuthal preferences were largely
decided by sightings upon the sun or moon. Somerville (1923, 215)
thought that the Hebridean tomb of Barpa Langass had been aligned on the
midwinter sunrise. Fragile threads of folklore may hold memories of this
association between the dead and the sky. The Waterstone tomb in
Somerset is said to dance on midsummer's day when there is a full moon
(Bord J. & C., 1978, 144; Grinsell, 1971, 68, 88), and in many Irish
tombs with carved stones there are symbols that appear to represent the
sun.

The need for a tomb to look towards the sun or moon,
however, was possibly only one of several factors that decided its
orientation and it is unlikely that many solar or lunar alignments were
precise. The configuration of the ridge or hillside on which the stones
were put up, the source of stone and its quality as building material,

the proximity to water or a settlement, the need to make the monument
conspicuous, any or all of these considerations could affect the
direction that the entrance would face. Nor would every group of people
have the skill or patience to attain the accuracy they wanted. When one
adds to these problems the fact that the passage, often not exactly
straight, might be a metre or more across but only three or four metres
long it will be realised that the arc of skyline visible from the
chamber will often be wide. Such a broad orientation quite possibly
satisfied the symbolic requirements of a Neolithic family but it will
hardly commend itself to a modern investigator looking for an unequivocal
sightline.

Gradually, over the centuries, these burial-places lost their
role as simple tombs and became ossuaries in which the skulls and
longbones of ancestors were used in rituals that involved the sun or
moon. Newly-built 'tombs' were sometimes grandiose structures with more
obvious astronomical properties. The roofbox at Newgrange, a passage-
grave erected around 3250 BC over a thousand years after the beginning
of the megalithic tradition, is an example of this change.

So perhaps, is the analogous solar 'letterbox' at Maeshowe
(Burl, 1981b, 125), a cairn tentatively dated to 2675 BC (2185+65 bc;
2020+70 bc). It faces SW and one fact, little commented upon, is that
frequently in the final age of chambered tomb building the entrances
faced westwards rather than to the east as though the builders were
anticipating the beliefs of the Iron Age Celts 2000 years later whose
ceremonies were nocturnal, the festivals starting at sunset. Of the few
dated late tombs Slieve Gullion, Co. Armagh, 2005+75 bc or 2525 BC,
faced SW, Cashtal yn Ard, Isle of Man, faced west, and the wedge-graves
of the west of Ireland had entrances consistently built between WNW and
SSW. This is a reversal of the easterly entrances of the earliest
Neolithic tombs and it implies a fundamental change of belief.

Increasingly formalised ceremonies took place outside the
entrances which were often elaborated by the addition of imposing
pillars to form deep semi-circular forecourts (Scott, 1969, 189). It is
likely that the splendidly impressive facades at Cairnholy I,
Kirkcudbright; King Harry's Grave, Isle of Man; Ballymacdermot,
Co. Armagh; the Bridestones, Staffs.; Carn Ban on Arran and many others
were the megalithic settings for rites held at the solstices or other

times of the year whose occasions were determined by the positions of
the sun or moon.

Within these U-shaped forecourts instead of a single family
many people could gather. It is impossible for us to know what part
astronomy played in their funerary rituals but the forecourts themselves
were the forerunners of stone circles in which the orientations are
unequivocal. At Meayll Hill on the Isle of Man (Kermode & Herdman, 1894)
the transformation was complete. Instead of the chambers of a tomb
being enclosed under a mound they were arranged as six T-shaped cists
in a segmented circle around an open central space. Significantly, the
two entrances to this sacred ring were at north and south, cardinal
positions which anticipated those of the stone circles of Cumbria
120 km (75 miles) to the NE across the Irish Sea.

Nowhere is this change from chambered tomb to stone circle
more manifest than in the development of the Scottish recumbent stone
circles from the Neolithic Clava Cairns of Inverness-shire. The unique
architectural details of these passage-graves with surrounding stone
circles have been fully described (Henshall, 1963, 358-91). More
recently, the astronomical information obtained from an analysis of their
declinations encourages the supposition that other groups of Neolithic
tombs will repay detailed archaeo-astronomical study.

Although there is no direct date two C-14 assays from a
pre-tomb phase at Raigmore suggest that the Clava Cairns belong quite
late in the Neolithic, the earliest not before 4000 BC. What is of
interest here (Fig. 1, A) is the clustering of the passage-grave
entrances in a narrow arc between 172° and 232° (Burl, 1981a, 258), the
azimuths of six of the eleven tombs available for examination being
taken from the accurate plans of Thom (Thom et al., 1980, 247-73).
Similar clustering can be seen in the related recumbent stone circles of
Scotland and Ireland (Fig. 1, B-D) showing how this tradition endured
for almost 3000 years.

Calculations of the declinations of the Clava passage-graves,
although not yet tested in the field, apparently reveal a definite
pattern of solar and lunar alignments towards the southern extremes, the
moon being more popular than the sun (Burl, 1981a, 260). Because no
passage is less than 50 cm (1 ft 6 ins) wide nor longer than 9 m (29 ft)

Fig.1.

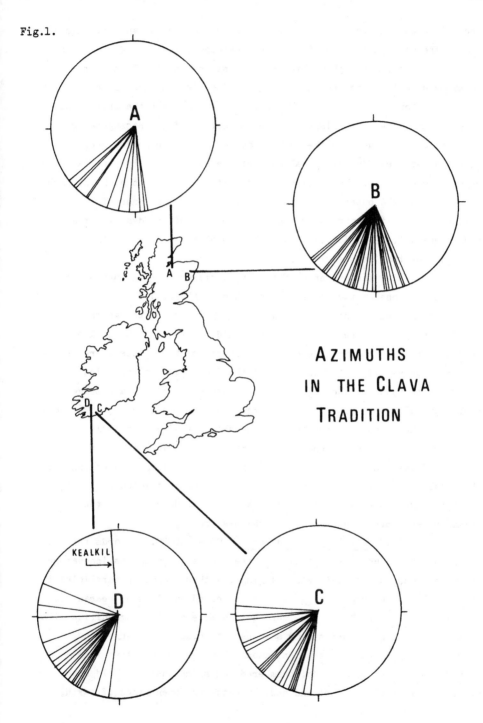

AZIMUTHS
IN THE CLAVA
TRADITION

there is always quite a wide visual arc from the chamber and for this
reason the declinations in Fig. 2 are presented as bars.

Yet, although the tombs were astronomically planned, the low
passages usually prevented any living observer from seeing either the
moon or sun outside the tomb. We must assume that once the appropriate
direction had been established and the sideslabs of the passage aligned
on it this was enough and the passage and chamber were roofed over.
There was no need for the actual rays of the sun or moon to penetrate the
megalithic darkness. This is in keeping with the cupmarks at Tordarroch,
carved not on the outer face of a kerbstone but on its inner side,
pressed against the cairn where only the spirits of the dead might see
them (Henshall, 1963, 385). It was the symbolism rather than the
physical act that was real.

This, the presence of quartz, cupmarks, and the absence of
any more than one or two bodies in the chamber (Burl, 1981b, 92)
reaffirms the belief that these cairns were shrines. Although the
sightlines in them were more accurately laid out than in the earliest
Neolithic tombs the cairns were not observatories but places where people
gathered for ritual. It could be that the Clava ring-cairns which are
mostly to be found to the east of the concentration of Clava passage-
graves and with much wider and uncovered central spaces (Burl, 1976, 162)
were a northern Scottish response to the need for bigger assembly-
enclosures much as the extended forecourts had been in some other
regions of the British Isles. Certainly it is the Clava ring-cairns with
their tallest circle-stones and kerbs in the same restricted arc as the
passage-grave entrances that act as one of the bridges between the
Primitive and Developed stages of archaeo-astronomy.

(B).2 Developed alignments. 3000-1800 BC. Towards the end
of the Neolithic and the beginning of the Bronze Age, with its innovatory
gold and bronzework, society was changing and this change is paralleled in
the ritual monuments of the time. The dispersed families of peasants,
each with its own tomb/ossuary, merged into more complex communities with
headmen and, perhaps, priests or shamans. Although the change was
gradual there is general agreement amongst archaeologists that from
around 3000 BC the building of chambered tombs declined, the old mounds
one by one were blocked up, and new, spacious and open rings of earth,

Fig.2

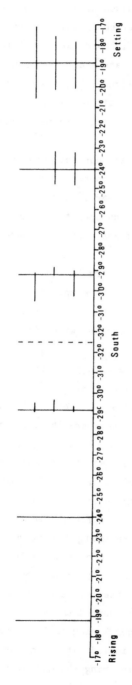

DECLINATIONS OF CLAVA PASSAGE-GRAVES (Provisional)

The vertical lines represent the declination of the sun at midwinter (-24.0), and of the moon at the major standstill (-29.2) and the minor standstill (-18.8).

timber or stone were erected.

Sometimes known as the Meldon Bridge phase after the huge
enclosure in Peebles-shire it is a period in which 'society was
remarkably well organised, with powerful and capable élites' (Burgess,
1980, 45). Despite the continuing debate about the nature of such a
development there are indisputable 'signs of increased centralisation
and ranking in social organisation' (Whittle, 1981, 320), revealed in
the construction of massive earthworks, settlements, henges and stone
circles.

If the earlier arguments are correct about the origins of
these circles being in the enlarged forecourts of chambered tombs then
four things could be expected of these early rings:-

a. They should be large and open to accommodate many people
because this was the reason for the transition from cramped tomb to
uncovered, bigger forecourt.

b. They would probably be circular just as many forecourts
were semi-circular because the circle is a simple and easily designed
shape.

c. They would have associations with human bone because
their ancestry lay with the tombs.

d. Because of their links with the oriented tombs their
design would include astronomical alignments.

These are the very features discovered in early rings. They
are large and circular (Burl, 1973), and where thorough excavations have
taken place cremated human bone has often been found in them (Burl, 1976,
93). In one of the earliest yet known, at Stenness in Orkney with two
C-14 assays of 2356±65 bc and 2238±70 bc or about 2970 BC, fragments of
human hand and legbones lay in the rock-cut ditch (Ritchie, 1976, 12).

To ensure the validity of the fourth requirement, the
alignments, strict criteria must be applied. The alignments should be
both inherent and demonstrable components of the circle. Features which
cannot be proved to be contemporary with the ring should be rejected.
Outlying stones are suspect. Unindicated skyline notches are almost
criminal. As an exercise the writer counted the dips, nicks and slopes
of the horizon around the Machrie Moor X ring on Arran and discovered no
fewer than 63 not one of which was singled out by any feature in the
circle itself. The fact that from the ring the midsummer sunrise happens

to coincide with the gap of the major pass through the Arran hills may
be no more than that, an inevitable consequence of the geographical
relationship between the pass and the once-fertile triangle of Machrie
Moor.

The stones of any circle should be their own record.
Entrances, tallest stones, graded heights, recumbent stones, the axis of
an ellipse, all these can reasonably be supposed original features and
if through them some astronomical event is defined this can be accepted
as significant, particularly if it is an occurrence repeated in several
similar sites.

The entrances of several south-western henges, for example,
are N-S aligned (Burl, 1970, 10) whereas henges in Wessex - Stonehenge,
Woodhenge, Maumbury Rings amongst them - had entrances to the NE,
perhaps towards midsummer sunrise. The long axis of the sophisticated
ellipse of stones at Cultoon on Islay was apparently designed to
indicate midwinter sunset over the distant Irish mountains (MacKie, 1981,
128).

The noticeably taller stones in certain regional groups of
circles also seem to occupy too narrow an arc for this to be accidental.
In Cumbria it is often a cardinal position that is so marked. Tall
stones at N, S, E or W are found at: Elva Plain (W), Brats Hill (S),
Swinside (N); and there are northern entrances at the Carles and
Gunnerkeld. Two massive stones stand almost E-W at Long Meg & Her
Daughters, possibly revealing how this major axis was laid out to the
midpoint between the midsummer and midwinter sunsets. In theory this
should have been an E-W line but at Long Meg the builders failed to
locate true west because the western skyline there has a pronounced slope
so that bisecting the extremes of the sunsets resulted in a slightly
ENE-WSW alignment.

Cardinal points have also been noticed at Stonehenge,
Brodgar, Stenness, Callanish, Kealkil, Druids Temple and others and it is
possible that research will show that many rings were set out in
quadrants determined by these positions, a suggestion which, if true,
could tell a great deal about the cosmological thinking of people in
their funerary rites 5000 years ago.

> At the round world's imagin'd corners, blow
> Your trumpets, Angels, and arise, arise

> From death, you numberless infinities
> Of souls ...
>
> John Donne. Holy Sonnets, vii

Talking of the American Plains Indians' Sun Dance, Collier (1947, 136) wrote, 'Besides the sun, other powers of the earth and sky, the thunder, the stars, mother earth, and the four cardinal directions, were represented in song, dance and painting', all this in the long ceremony that unified the tribe, 'the culminating discipline, forthgiving and sharing, which structured the personality of the young, renewed the personality of the old, opened the mind's windows to a noble world-view' (ibid, 135).

The Sun Dance rituals inside an east-facing lodge held little suggestion of any scientific astronomy but the significance of the cardinal points and their symbolic integration in the Indian vision of the world is clear. It may be that some British stone circles also were raised by people with a comparable picture of their universe.

There was something of a duality in the planning of the Cumbrian rings. Although the cardinal positions were marked, often another direction also was established, a 'secondary' alignment that may have been used calendrically to define the occasion when a ceremony should be held. At Long Meg there is a SW entrance in line with midwinter sunset; a SE entrance at Swinside; a tall radial stone at the Carles from which midsummer sunset could be seen (Thom, 1967, 99); tall N and S stones at Ballynoe, Co. Down, but an entrance at the WSW.

Such WSW alignments are of interest because they are a feature of many other western circles. The two Llandegai henges had WSW entrances. So did the Druids' Circle on Penmaenmawr and the Stripple Stones circle-henge on Bodmin Moor. Many Cornish rings have their tallest stone there (Tregelles, 1906) and at Boscawen-Un the only quartz stone in the ring stands at the WSW. At the Lios, a circle-henge in Co. Limerick, Windle (1912, 287) calculated that two massive stones arranged like a gunsight at the WSW opposite the imposing entrance were in line with sunset at the beginning of November, the time of the Celtic Samain festival when the dead were supposed to rise from their graves. A somewhat similar bearing can be seen at Beltany Tops, an evocatively-named circle in Co. Donegal, where from a copiously-cupmarked stone at the ENE one can look across the ring to the tallest stone, a 3 m (9 ft 6 ins) high pillar at the WSW (Somerville, 1923, 213), the azimuth coinciding with sunset at the end of October.

If it could be demonstrated by analysis of a group of these WSW alignments that the builders had been setting up calendrical sight-lines this would contribute enormously to our understanding of prehistoric ritual and the origins of such 'Celtic' festivals as Samain in late October. If the druids had a Bronze Age ancestry (MacKie, 1981, 141) there is no reason why 'Iron Age' festivals should not be just as old (Burl, 1979, 204).

Perhaps the most subtle way of indicating an orientation was by grading the heights of the stones, most elegantly achieved by the choice of stone-length and measured depth of each stonehole. It was a widespread practice and not to be confused with the coarser custom of erecting two taller stones in a ring like tower-blocks in a residential suburb. In the SW the rings of Gors Fawr in Dyfed and Brisworthy and Fernworthy in Devon have stones gently graded in height. The five Stonehenge trilithons were graded upwards to the SW and the comparable henge of Lugg in Co. Dublin had three sets of paired posts, each perhaps with a surmounting lintel, that grew in width and height towards a massive post at the W (Kilbride-Jones, 1950, 324). There are other examples of grading and the tradition is seen at its most effective in the recumbent stone circles of NE Scotland, those dramatic rings whose architecture was derived from the Clava Cairns of Inverness-shire.

The architecture and contents of these circles have been described (Burl, 1976, 167) and their astronomical azimuths have been tentatively explained (Burl, 1980). The presence of cremated human bone, quartz, cupmarks and graded circles link them exclusively with the Clava Cairns but the earliest of these rings was built much later than the first of the passage-graves, probably within a century or two of 2500 BC. It is the recumbent stone, a huge block laid flat between the two tallest stones, that distinguishes these circles from the Clava sites in which there is no such feature. It is noticeable, however, that the azimuths of the recumbent stone circles (Fig. 1, B) occupy an arc almost identical with that of the passage-grave entrances. They have been interpreted as attempts to align the recumbent stone on the extreme of the southern moon between its rising and setting, and a provisional analysis of the declinations from 35 sites (Fig. 3) offers some confirmation of this.

These are not finely-defined alignments. In an average circle

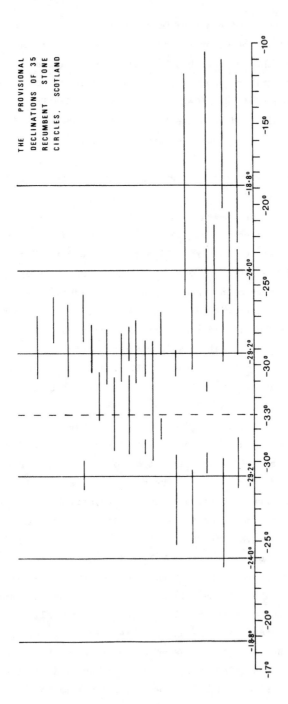

THE PROVISIONAL
DECLINATIONS OF 35
RECUMBENT STONE
CIRCLES. SCOTLAND

Fig.3

with a diameter of 20 m (66 ft) and a recumbent stone 3.7 m (12 ft) long
the arc of vision between the flankers from the other side of the ring
would be over 10° wide. For this reason the declinations are shown as
bars whose ends represent the east and west flankers. Nevertheless, the
diagram does show a close correlation between extreme southern moonrise
(5 sites), southern moonset (14 sites) and minimum southern moonset (4
sites) with many other rings directed to the moon at the south when it
was well clear of the horizon. Conversely, there seems to have been
little interest in the midwinter sunset (5 possible sites).

It must be repeated that this is a provisional interpretation.
Dr. Clive Ruggles is preparing the definitive report but even in the
elementary form shown here the declinations do appear to reveal the strong
interest the builders had in the moon.

This is in keeping with the evidence from the other circles,
of good but broad sightlines towards the sun or moon in rings where the
living thought it necessary to have some association with the dead. We
do not yet know why. What we can say is that there are apparent
alignments, that the circles are well-designed enclosures and that their
size suggests that they were intended for a considerable number of people.
Quite possibly, as MacKie (1981, 141) has suggested, these people
included astronomer-priests - or shamans or medicine men or witchdoctors,
the terms being interchangeable according to one's anthropological
prejudices. In succeeding periods, however, there was a falling-off in
the spaciousness and design of the rings still to be built. The custom
of planning and erecting ritual enclosures slowly declined into the tiny
circles and the last, sad stones of the megalithic tradition.

(B). 3. Local alignments. 1800-1300 BC. In the Overton and
Bedd Branwen phases of what were once called the Early and Middle Bronze
Ages there was increasing settlement on hill-farming country by people
moving out of the overpopulated lowlands (Burgess, 1980, 118). The
milder climate permitted the cultivation of the poorer, colder soils of
the uplands and it was during this period that many of the smaller stone
monuments on marginal lands such as Dartmoor and the Cheviots were built.

In landscapes of hills, valleys and splashing rivers isolated
homesteads were customary, each family almost independent of the others,
and there was probably a looser social hierarchy than that which
persisted in long-settled regions. Whether in the Sperrin mountains of

Ulster or on the Yorkshire Moors or elsewhere the new ritual centres were
small, sometimes rather crudely constructed by a few men and women who
had no need of a large enclosure. Nor did they build exact copies of the
circles of their forefathers. Instead, one finds approximations, the
4-Posters of central Scotland, the cairn-circles of Dartmoor,
unspectacular sites that can still be seen in the gorse and heather of
marginal farming land like the millstone grit of the Peak District
(Hawke-Smith, 1981, 62) with its ring-cairns, kerb-circles and enclosed
cremation cemeteries. There are so many of these inconspicuous 'family'
rings in the British Isles that, in one sense, the Thoms (A. & A. S.,
1978, 178) were right when they claimed that every small community had
its own observatory. It must be added, however, that only too often the
astronomical mechanism of such sites was clumsy and careless.

    Only a few people could have used such mini-megaliths.
Whereas an ordinary stone circle of the Developed phase is 20 m (66 ft)
or more in diameter the moorland rings of Dartmoor on average are only
10 m (33 ft) across. In Co. Londonderry the circles of unobtrusive
stones are similar, 10.6 m (34 ft) wide. The longer axis of a 4-Poster
in Scotland or a related 5-Stone ring of SW Ireland is only 4.9 m (16 ft)
and 3.2 m (10 ft 6 ins) respectively. Even Callanish, the 'Stonehenge
of the North', measures no more than 13.1 x 11.3 m (43 x 37 ft), showing
that, whatever its function, not many people could have participated in
rituals inside the circle.

    Variations of megalithic monuments are also to be found in
this Local phase. There are some horseshoe-shaped settings of standing
stones (Burl, 1976, 157), often open to the SW, and rows of stones whose
date is not known directly but whose presence 'consistently in the
moorland fringes' (Emmett, 1979, 109), whether in Caithness, western
Scotland or in central Wales, suggests that they also were the products
of people at this time. Supporting evidence comes from a C-14 assay of
1315±55 bc (c. 1610 BC) for the alignment at Maughansilly, Co. Kerry
(Lynch, A., 1980, 73) and from a less certain one of 1380±50 bc (c. 1685
BC) from the Dromatouk row in the same region.

    Whether row or ring, what one does find in many of them is
human bone. Cremations were buried by the pair of stones at Orwell,
Perthshire (Ritchie, 1974, 8), a site that remained a cult centre for
years. Urned cremations lay inside the stone ring of Barbrook II,

Derbyshire, dated to about 1800 BC. Burnt bone was discovered in the
large cairn at Kintraw, Argyll, whose 'blind entrance' at the SW passes
sadly undiscussed by advocates of the astronomical interpretation of this
site despite the existence of comparable SW 'entrances' in other cairns
(Burl, 1976, 171). Even the now empty cist and dismantled cairn at
Ballochroy hint that the builders may have regarded the stone row there
as an elemental version of a chambered tomb, the stones representing the
backslabs of a row of chambers as they would have appeared before a long
cairn was heaped over them. If this was the significance of some stone
rows it might explain their distribution in regions of megalithic tombs
(Emmett, 1979, 108).

        Such associations between megaliths and death are commonplace
but the tininess of the rings and the shortness of many rows precluded
any neat sightline being preserved in them. It is unlikely, moreover,
that the families using them needed anything more than a general
orientation to the sun or moon. Again, it is the Clava tradition, even
in its final stages, that provides some confirmation of this.

        The architecture of the recumbent stone circles (RSCs) of
SW Ireland is not identical to that of the Scottish rings but there are
so many similarities that a direct connection is likely. Whether this
involved a movement of people is unknown although the idiosyncratic
recumbent stone circle of Torhousekie in Wigtownshire, perhaps aligned
on the midwinter sunrise, does lie on the route between NE Scotland and
Munster. A 'Scottish' 4-Poster at the Four Stones, Walton, Powys and a
comparable 'Irish' 5-Stone ring at Circle 275, Penmaenmawr, Gwynedd
also occur in intermediate locations.

        In the counties of Cork and Kerry the circles that most
nearly resemble the Scottish RSCs lie in coastal situations. They are
not copies. They are smaller and their tallest pillars stand not along-
side but opposite the recumbent stone which itself is often a shortish
slab rather than a heavy block. It is as though several generations had
passed before an incoming family had been able to erect a stone circle.
By then there would have been no one alive who had ever seen a Scottish
ring.

        If the builders of the Irish circles were the descendants of
immigrants from Scotland the tedious migration southwards may have taken

many years. More years still would have be_n spent clearing land, building a home, establishing crops and cattle before time could be spared to
erect a family's permanent shrine. Remembering always that we are
discussing people and not merely a mechanical Megalithic Man it is to be
expected that over such a long and demanding period the inherited image
of an ancestral stone circle should have become distorted. Indeed, it is
remarkable that so much of the architectural framework - the recumbent,
the grading, the quartz, the cupmarks, the cremated bone, the
orientation to the SW quadrant - was retained.

        The same process of transmutation occurred further inland in
the hills of the Boggeraghs. Here the design of the even smaller 5-Stone
rings with their diminutive recumbents and four circle-stones recalls
the 4-Poster rings of central Scotland but, again, in a modified form.
When to these changes are added the effects of other traditions such as
the idea of a central pillar, known in at least ten of the Irish rings
and possibly the result of Cornish influence, it is clear that the
interpretation of archaeo-astronomical practices will be very dependent
upon some understanding of the archaeology of these late sites. It has
been supposed that these Irish circles are late in the megalithic
tradition (Burl, 1976, 223) and recent excavations support this. Two
C-14 dates of 970±60 bc and 715±50 bc show that Cashelkeelty 5-Stone
ring in Co. Kerry was put up between 1200 and 900 BC (Lynch, A., 1980,
66) almost 3000 years after the earliest of the Clava tombs.

        It is not only the architecture but the astronomy, of course,
that justifies the Clava-Aberdeen-Cork/Kerry connection. The azimuths
of both the coastal Irish circles (Fig. 1, C) and the inland 5-Stone
rings (Fig. 1, D) cover exactly the same SSE-SW arc as the Scottish
chambered tombs and RSCs so that in monuments of such similar design
there can be some confidence that they were culturally related. In the
Irish sites, however, there are further azimuths around WSW and even WNW.
A simple archaeological explanation can be provided for these.

        Various astronomical studies have been made of the Irish RSCs.
Somerville (1930) noticed that five he had inspected all stood on
summits or south-facing slopes exposed to the sun. Fahy (1962), although
previously satisfied with the sunset associations at Bohonagh and
Drombeg, concluded at Reanascreena that there was no significant
orientation there.

Barber (1973) analysed the alignments in 30 rings and despite some inconsistency in the results (Freeman & Elmore, 1979, 90) it is interesting to find that 20 of his declinations, those with azimuths between 171° and 235°, conform to the lunar hypothesis proposed for the Scottish sites. Gortanimill with its recumbent stone at 171° is a fine circle 8.1 m (27 ft) in diameter, with a squat quartz stone near its centre. An axial line to the recumbent's midpoint produced a declination of -35.5°, maybe towards the moon high in the sky. But so widely spaced are the stones on either side of the metre-long recumbent that an observer would have had an arc of vision between them over 22° wide.

Of the azimuths towards the SW, Oughtihery, a small ring only 3.3 m (11 ft) across, faced 230° with an estimated declination of -18.8° towards the minimum setting of the moon. In this compact circle the recumbent measures a disproportionately long 1.9 m (6 ft), giving an arc of vision of nearly 33° from SSW to WSW, a visual splay quite different from the delicate sightline required for a priestly observatory.

Of the ten remaining sites unaccounted for by the lunar theory one, Kealkil, has a northerly azimuth of 356°. The nine other rings have azimuths between 237° and 274°. Reanascreena (258°) and Bohonagh (265°), both excavated by Fahy, can be added to Barber's (1973, 32) list. These eleven anomalous rings have a significant distribution.

The classic 'lunar' sites such as Drombeg (226°), Templebryan (199°) and Ardgroom (195°) are coastal whereas the majority of the eleven 'non-lunar' rings are further inland amongst the hills. Five are in the Boggeraghs (Belmount; Carrigagulla A & B; Cousane; Coolclevane), two (Derreenataggart; Dromroe) are on the Slieve Mishkish peninsula north of Bantry Bay, and three (Reanascreena; Knocks; Maulatanvally) are in the hills around Carrigfadda. Only Bohonagh is on the coast, (Fig. 4).

The unusual portal-stones at right-angles to the circumference at Bohonagh, similar 'radial' portals and a quartz centre-stone at Maulatanvally, and the surrounding hengelike ditch and bank at Reanascreena suggest that all three circles may belong to the later stages of the Local phase in SW Ireland.

With the exception of Bohonagh the other rings with untypical azimuths were raised in the hills where chambered tombs known

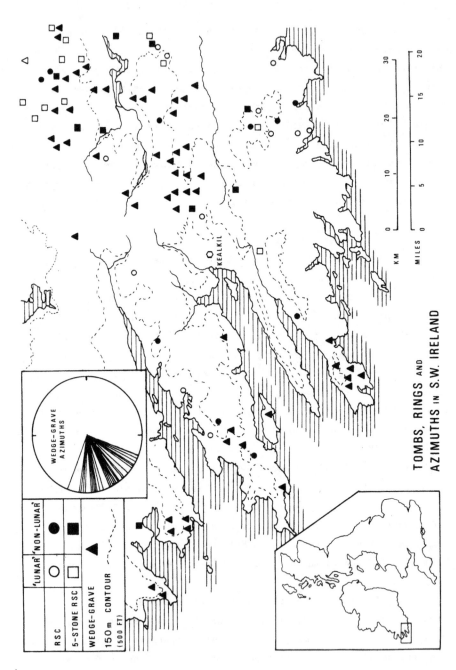

Fig. 4.

TOMBS, RINGS AND
AZIMUTHS IN S.W. IRELAND

KEALKIL

as wedge-graves already existed (O'Nuallain, 1975, 100). Some 70 of
these wedge-shaped megalithic monuments are known in Cork and Kerry,
their cairns rising in height towards the entrances which customarily lay
between SW and WNW. In SW Ireland it seems that 'the first settlers were
the wedge-tomb builders who arrived some time in late Neolithic or early
Bronze Age times' (Lynch, A., 1980, 17), establishing themselves around
the headwaters of the Lee and Brandon rivers and on the fingerlike
peninsulas of Kerry, in the hills north of Bantry exactly where the
stone circles with 'aberrant' azimuths are to be found. It is feasible,
therefore, that these extra orientations to the WSW resulted from the
mingling of the newcomers with natives whose traditions were unalterably
fixed in the tombs and alignments of stones that had been put up long
before the arrival of the circle-builders. At Cashelkeelty an E-W row
of four stones was in existence before the 5-Stone ring there (Lynch, A.,
1980, 65-6).

     The WSW bearings of the wedge-graves and of some of the Irish
RSCs may have been directed towards the November sunset, anticipating
the Iron Age festival of Samain. The WSW associations of many Cornish
rings is also of interest because in them the presence of quartz pillars
at Duloe and Boscawen-Un and the existence of central stones like others
in Cork and Kerry are features that may reveal links between the English
SW peninsula and the copper-producing lands of SW Ireland during the
2nd millennium BC. Again, without some knowledge not only of the stone
circles but of the wedge-graves as a group the interpretation of such
astronomical nuances would not be possible. It shows how necessary the
co-operation of all disciplines is in the pursuit of these archaeoastronomical mysteries.

     Yet, whether the moon of the RSCs or the setting sun of the
wedge-graves the astronomy was fused with death. At the centre of the
Cashelkeelty ring was a slab-covered pit and in it were the cremated
remains of a young adult. Death was the constant companion of these
stones.

     (C) <u>Pi in the sky</u>. Much reliance has quite properly been
placed on the ability of statistical analysis to demonstrate how strong
any claim for an alignment is. With researchers involved in the attempt
to improve such analyses and with the current belief that in future more
exacting criteria will be adopted by archaeo-astronomers one can be

optimistic about the ultimate identification of alignments in megalithic
monuments.

This, however, is only the first move in a three-dimensional
game. Statistical methods may enable us to recognise an alignment but
this is all they will do. They may tell us 'what happened', and
archaeologists with some knowledge of the design of tombs and enclosures
may tell us 'how it happened' but neither, as the Greeks discovered, will
tell us what we really want to know, 'why it happened'. For this we may
have to depend upon anthropology, a discipline as alluring as the golden-
voiced Sirens ... and, perhaps, in the British Isles, as treacherous.

The problem is three-dimensional because it involves not only
the detection of alignments but also the study of structures widely
distributed from the Shetlands to Brittany and from eastern England to
the western shores of Ireland. These structures, moreover, had different
functions. They were also lengthily separated in time.

These are factors that have been usually ignored in
astronomical works but it is essential for us to understand that during
3000 years of prehistory beliefs and practices must have changed. Simply
because the Thoms in their pioneering fieldwork have generally
concentrated upon the monuments of the Local phase this does not absolve
researchers from recognising the probability that those rough rows and
shrunken circles were only the last vestiges of a long tradition, the
handiwork of peasant families whose more casual alignments were often less
exact but no less important to their creators than the finer alignments
of the Developed phase. Priests, possibly, at Stonehenge. Peasants,
probably, at Ballochroy. Even in a single area the chambered tomb, the
megalithic circle and the row of standing stones may represent over a
hundred generations of cultural change, as distant from each other as we
are from David and Goliath and from the Minoan palaces of Crete.

From early chambered tombs such as Ascott-under-Wychwood in
Oxfordshire, built around 3800 BC, down to the tiny ring of posts at
Itford, in Sussex, put up by an impoverished family around 1250 BC,
stretch almost three millennia of changing social conditions, changing
climate, changing economy and changing ritual practices. Even within the
British Isles each region is different. There are so many variables that
it is little more than vapid to remark that Newgrange, Stonehenge and
Ballochroy contain solar alignments. A little more substance is added by

the observation that all three sites were also associated with death. One
supposition might be that in some way the sun or moon were believed to
hold the promise, after death, of an easier, more comfortable life:

> You will eat, bye and bye,
> In that glorious land above the sky.
> Work and pray, live on hay,
> You'll get pie in the sky, when you die .
> Joe Hill. The Little Red Song Book.

It will be for anthropologists, from their knowledge of other
societies, to tell us whether it was the expectation of some glorious
solar heaven or luxurious lunar paradise that motivated people in the
prehistoric British Isles. One suspects, though, that the possible
reasons for the alignments are so many that any attempt at nomothetic
interpretation will be revealed as no more than a superficial
generalisation. More material is needed, more study of other aspects of
prehistoric religion and ritual, more knowledge of the life and times of
the people whose ancient shrines we try to decipher. Comfortingly, there
are two approaches that seem to offer some chance of success.

In the three stages of astronomical development proposed in
this paper there are some obvious, clearcut groups for the enthusiast to
study. In the Primitive phase there are the aggregations of chambered
tombs such as the Cotswold-Severn group; the Clyde-Solway group; the
Camster tombs. In the Developed phase one might look at the splendid
Cumbrian stone circles to see if there was a sustained interest in the
cardinal points. In the Local phase there are the multiple rows of
northern Scotland now helpfully listed by Myatt (1980), or the pairs of
standing stones in Perthshire (Stewart, 1967, 142-5) on which very little
astronomical work has been done.

One might also look at a particular region to see whether
there had been an astronomical tradition that endured from the Primitive
to the Local phase. As an example, one might consider the monuments
around the Preseli mountains of Dyfed, the source of Stonehenge's disputed
bluestones. Here there is Pentre-yfan, a portal-dolmen facing southwards
and built over an earlier, E-W arranged standing stone. There is Bedd-yr-
afanc long chambered tomb whose mound extends E-W. Of a later period
there was Cerrig-y-gof, a cist-circle somewhat akin to Meayll Hill on the
Isle of Man, with a gap in its ring almost at the south, and there is the
fine stone circle of Gors Fawr, its stones graded to the south but with

two standing stones to the NNE. Their axis is SW-NE, possibly pointing
to the midsummer sunrise (Thom, 1967, 101, W9/2). And on the other side
of the mountains there is the ruined line of stones at Parc-y-meirw,
perhaps directed towards the minor northern setting of the moon (ibid,
W9/7).

Only assiduous fieldwork could tell whether a majority of
these monuments did contain alignments set approximately to the cardinal
points. Study of the NNE-SSW 'lines' in the Kilmartin Valley of Argyll
might be equally instructive. Together, such group and regional studies
might clarify many of the questions that still bedevil us. At present
what evidence we have supports a belief in ceremonial, even calendrical
sightlines. Cardinal directions may also be detected if whole groups of
monuments are considered. This could be especially true of stone circles.

> Where shall we find two better hemispheares,
> Without sharp North, without declining West?
> John Donne. The Good-morrow.

With persistent study, maybe nowhere. Like pencil-rubbed paper over a
penny unsuspected patterns may appear as our approaches improve. There
may well have been pi in our megalithic skies. And as we change courses
we may at last receive our long-awaited desserts.

Ashmore P. (1980). Callanish, 1980. Interim Report. Edinburgh.
    Scottish Development Department.
Barber J.W. (1973). The orientation of the recumbent-stone circles of
    the south-west of Ireland. Journal of the Kerry Archaeological &
    Historical Society 6, 26-39.
Barker G. (1981). ed. Prehistoric Communities in Northern England.
    Sheffield. Sheffield University.
Barlai K. (1980). On the orientation of graves in prehistoric
    cemeteries. Archaeoastronomy Bulletin III(4), 29-32
Bord J. & C. (1978). The Secret Country. St. Albans. Paladin.
Burgess C. (1980). The Age of Stonehenge. London. J.M. Dent.
Burl A. (1970). Henges: internal features & regional groups.
    Archaeological Journal 126, 1-28
    - (1973). Dating the British stone circles. American Scientist
    61 (2), 167-74
    - (1976). The Stone Circles of the British Isles. London & New Haven.
    Yale U.P.
    - (1979). Prehistoric Avebury. London & New Haven. Yale U.P.
    - (1980). Science or symbolism: problems of archaeo-astronomy.
    Antiquity 54, 191-200
    - (1981a). By the light of the cinerary moon. In: (eds) Ruggles &
    Whittle. 243-74.
    - (1981b). Rites of the Gods. London. J.M. Dent.
Collier D. (1947). Indians of the Americas. New York. Mentor.
Cunnington W. (1857). Account of a barrow on Roundway Hill near Devizes.
    Wiltshire Archaeological Magazine 3, 185-8.
Emmett D.D. (1979). Stone rows: the traditional view reconsidered. In:
    (ed). V.A. Maxfield. Prehistoric Dartmoor in its Context. Torquay.
    Devon Archaeological Society. 94-114.
Fahy E.M. (1962). A recumbent stone circle at Reanascreena South, Co.
    Cork. Journal of the Cork Historical & Archaeological Society 67,
    59-69.
Fichter G. & Volk P.J. (1980). The eastern orientation of Merovingian
    graves and the seasonal distribution of morbidity and mortality ...
    Journal of Human Evolution 9, 49-59.
Freeman P.R. & Elmore W. (1979). A test for the significance of
    astronomical alignments. Archaeoastronomy 1, 86-96.
Grinsell L.V. (1971). Somerset barrows, part 2. North & East. Proc.
    Somerset Archaeological & Natural History Society, 115, 44-137
Harrison R. (1980). The Beaker Folk. London. Thames & Hudson.
Hawke-Smith C.F. (1981). Land-use, burial practice and territories in
    the Peak District, c. 2000-1000 bc. In: (ed) Barker G. 57-72.
Henshall A.S. (1963). The Chambered Tombs of Scotland, I. Edinburgh.
    Edinburgh U.P.
Higgins G. (1829). The Celtic Druids. London.
Huntingdon R. & Metcalf P. (1979). Celebrations of Death. The
    Anthropology of Mortuary Ritual. Cambridge. Cambridge U.P.
Kalicz N. (1970). Clay Gods. The Neolithic Period and Copper Age in
    Hungary. Budapest. Corvina.
Kermode P.M.C. & Herdman W.A. (1894). The excavation of the Neolithic
    stone circle on the Meayll Hill ... Isle of Man. Trans. Liverpool
    Biological Society 8, 1-14.
Kilbride-Jones H.E. (1950). The excavation ... at Lugg, Co. Dublin.
    Proc. Royal Irish Academy 53C, 311-32.

Kinnes I., Chapman & Randsborg K. (1981). eds. The Archaeology of
    Death. Cambridge. Cambridge U.P.
Lynch A. (1980). Man & Environment in S.W. Ireland. 4000 BC - AD 800.
    Oxford. British Archaeological Reports.
Lynch F. (1973). The use of the passage in certain passage graves as a
    means of communication rather than access. In: (eds) Daniel G. &
    Kjaerum P. Megalithic Graves & Ritual. Jutland Archaeological
    Society Publication 11, 147-61.
Lockyer N. (1906). Stonehenge and Other British Stone Monuments
    Astronomically Considered. London. Macmillan.
  - (1909). ibid. 2nd edition.
MacKie E.W. (1981). Wise men in antiquity? In: (eds) Ruggles &
    Whittle. 111-52
Milisauskas S. (1978). European Prehistory. New York & London.
    Academic Press.
Mortimer J. (1905). Forty Years' Researches in ... the Burial Mounds of
    East Yorkshire. London. A. Brown.
Myatt L.J. (1980). The multiple stone rows of Caithness and Sutherland.
    Caithness Field Club Bulletin 2(7), 191-5.
O'Kelly M.J., Lynch F. & O'Kelly C. (1978). Three passage-graves at
    Newgrange, Co. Meath. PRIA 78C (10), 249-352.
O'Nuallain S.P. (1975). The stone circle complex of Cork & Kerry.
    Journal of the Royal Academy of Ireland 105, 83-131.
Ponting G. &. M. (1981). Decoding the Callanish complex - some initial
    results. In: (eds) Ruggles & Whittle. 63-110.
Powell T.G.E., Corcoran J.X.W.P., Lynch F. & Scott J.G. (1969).
    Megalithic Enquiries in the West of Britain. Liverpool. Liverpool
    U.P.
Ritchie J.N.G. (1974). Excavation of the stone circle and cairn at
    Balbirnie, Fife. Arch J 131, 1-32.
  - (1976). The stones of Stenness, Orkney. Proc. Society of
    Antiquaries of Scotland 107, 1-60.
Ruggles C.L.N. & Whittle A.W.R. (1981). eds. Astronomy & Society in
    Britain during the Period 4000-1500 BC. Oxford. British
    Archaeological Reports.
Scott J.G. (1969). The Clyde cairns of Scotland. In: Powell et al.
    175-222.
Simpson D.D.A. (1976). The Late Neolithic and Beaker settlement at
    Northton, Isle of Harris. In: (eds) Burgess C. & Miket R.
    Settlement & Economy in the 3rd and 2nd Millennia BC. Oxford.
    British Archaeological Reports. 221-31
Somerville H.B. (1913). Astronomical indications in the megalithic
    monument at Callanish. Journal of the British Archaeological
    Association 23 (1912), 83-96.
  - (1923). Instances of orientation in prehistoric monuments of the
    British Isles. Archaeologia 73, 193-224.
  - (1930). Five stone circles of west Cork. JCHAS 35, 70-85.
Stewart M.E.C. (1967). The excavation of a setting of standing stones
    at Lundin Farm. PSAS 98 (1964-6), 126-49.
Stukeley W. (1740). Stonehenge a Temple Restor'd to the British Druids.
    London.
Thom A. (1967). Megalithic Sites in Britain. Oxford. Oxford U.P.
  - & A.S. (1978). Megalithic Remains in Britain & Brittany. Oxford.
    Oxford U.P.

Thom A., A.S. & Burl A. (1980). Megalithic Rings: Plans & Data for 229
    Sites in Britain. Oxford. British Archaeological Reports.
Thurnam J. (1871). Ancient British barrows (part 2. Round barrows).
    Arch 43, 285-544.
Tratman E.K. (1976). A Late Upper Palaeolithic calculator (?), Gough's
    Cave, Cheddar, Somerset. Proc. University of Bristol
    Spelaeological Society 14 (2), 123-9.
Tregelles G.F. (1906). The stone circles. In: Victoria County History
    of Cornwall, I. London. Oxford U.P. 379-406.
Tuckwell A. (1975). Patterns of burial orientation in the round barrows
    of East Yorkshire. Bulletin of the Institute of Archaeology 12,
    95-123.
Turner A.W. (1976). Houses for the Dead. New York. McKay.
Whittle A.W.R. (1981). Later Neolithic society in Britain: a
    realignment. In: (eds) Ruggles & Whittle. 297-342.
Windle B.C.A. (1912). Megalithic remains surrounding Lough Gur.
    PRIA 30, 283-306.

A SURVEY OF THE BARBROOK STONE CIRCLES AND THEIR CLAIMED
ASTRONOMICAL ALIGNMENTS

R.P. Norris & P.N. Appleton
Nuffield Radio Astronomy Laboratories,
Jodrell Bank, Macclesfield, Cheshire

R.W. Few
147 Girton Road, Girton, Cambridge

Abstract   The Barbrook stone circles and their outliers have
been the subject of a total of 16 claimed astronomical alignments (Thom 1967, Barnatt 1978). We have re-surveyed the
sites in order to test these claims. In choosing alignments
we have adhered to rigidly defined selection criteria, and we
have applied the test of Freeman and Elmore, together with
pseudo-random simulations, in order to test the significance
of our results. We conclude that there is no evidence of any
deliberate accurate astronomical alignments, although there
is marginal evidence for rough astronomical alignments which
may have been constructed for ritual purposes.

Introduction

The valley of Barbrook on Big Moor in Derbyshire was the site
of a major prehistoric settlement (Radley 1966, Burl 1976). The moor
contains the remains of a number of stone circles, earthen enclosures,
standing stones and cairns, some of which have been shown by Lewis (1966)
to have existed in Neolithic and Early Bronze age times. One of the stone
circles, Barbrook 1, has been claimed by Thom (1967) and by Barnatt (1978)
to have been used by prehistoric man for astronomical purposes. In
addition, Thom (1967) has suggested that Barbrook 1 and another nearby
stone circle Barbrook 3 (also known as Owler Bar) were constructed
according to an accurate geometrical method (Thom type B circles).

In order to make an objective assessment of the
astronomical and geometrical claims two projects were
carried out:
(i) We have accurately surveyed possible sightlines following the criteria suggested by Cooke et al.(1977)
and have assessed their significance.
(ii) We have made accurate ground plans of the two
relatively complete stone circles Barbrook 1 and 3.

In the case of (i) the indicated astronomical declinations derived from
the alignments have been tested for astronomical significance using the

statistical method of Freeman and Elmore (1979).

The Barbrook Stone Circles and the extent of the Survey

In the Barbrook valley are a number of earthen ringworks, stone circles and cairns which were probably constructed around 1500 B.C. (Lewis 1966). Although Radley (1966) stresses the similarity between the earthen ringworks and the stone circles, we have restricted the survey to the stone circles and standing stones because it is around these that the existing astronomical hypotheses have been constructed. Furthermore, stone circles and standing stones comprise a well-defined group, allowing statistically complete surveys and minimising selection effects due to personal bias.

Fig. 1 is a map of the area showing the positions of the stone circles included in this survey, and for which we have adopted the numbering scheme of Barnatt (1978). Two additional nearby stone circles, Barbrook 4 and Stoke Flat, have been omitted from the survey because they have no outliers and are not visible from any of the other circles.

Fig. 1 The location of the four circles (indicated by numbers) included in the survey.

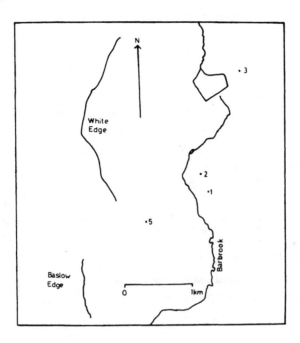

The Outliers at Barbrook 1

Fig. 2 shows the positions of the outliers at Barbrook 1. Barnatt lists 9 outliers, but we consider that, in the absence of definite archaeological evidence, not all of these are necessarily prehistoric standing stones. However, we have included them all for the purpose of testing his claimed alignments. Stones I,G,H,E,F and A all appear to be authentic. B and C may be standing stones but could be natural boulders, and D is part of the structure of a ruined cairn. Stone I is a fine cup-marked standing stone, but is not visible from Barbrook I and cannot therefore be called an outlier. We have therefore classified it as a separate site, and since it is visible from stones G and H, this leads to four extra possible alignments, although one of these is eliminated by the criteria discussed below.

Surveying Methods

The sites were surveyed on several separate occasions using a Watts one arcsecond theodolite, and steel or glass fibre measuring tapes. The direction of due North was accurately determined by observations of

Fig.2   The location of possible outliers near Barbrook 1. The numbers against each outlier give the height (in m) of the ground at each stone relative to the ground level in the centre of the circle.

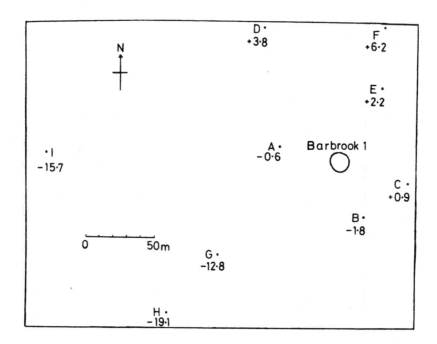

the Sun, timed using a calibrated quartz crystal wristwatch.   Several points
on each indicated horizon profile were surveyed, and later superimposed on
a photograph taken from the theodolite position using a telephoto lens.
The accuracy of these surveys is estimated to be better than one arc minute.   For the inter-site surveys, the distant site was marked by a ranging
pole.   The theodolite position when surveying from a circle was not
necessarily the centre of the circle, so azimuths and elevations were
subsequently corrected for the effects of parallax.   Elevations have also
been corrected for the effects of refraction.

     The ground plan survey of Barbrook 1 (Fig.3) was made using
the method described by Ruggles and Norris (1980):   the position of a
point on each stone was measured using the theodolite and measuring tape,
and then the orientation and dimensions of each stone were measured using
a magnetic compass and measuring tape.   The uncertainty in the position
of any point on a store is estimated to be less than 0.04m.

     The ground plan survey of Barbrook 3(Fig.4) was made using a
method in which distances between points on each stone were measured

     Fig. 3   Plan of the Barbrook 1 stone circle, reduced to a
horizontal plane.   Broken lines indicate an uncertain edge
due to vegetation

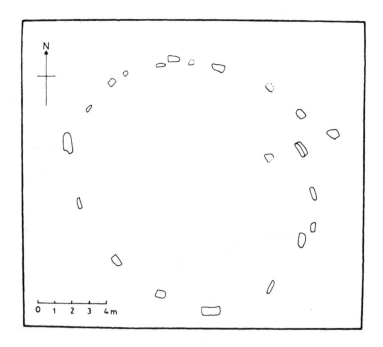

using a steel measuring tape. Many more such measurements were made than
were required to calculate the relative positions of the stones. The data
were then reduced by an iterative least squares fitting computer program,
which produces not only a plan of the stones but also an estimate of the
errors. The orientation and dimensions of each stone were again measured
using a magnetic compass and steel measuring tape.

This latter method has the advantage of speed and removes the
need to carry a theodolite to remote sites. It produces similar uncertainties to those produced by the method used at Barbrook 1, but does
not, of course, give an accurate orientation relative to true North. As
a check, this method was later applied to Barbrook 1:   the results of the
two surveys agreed within the uncertainties.

Alignments chosen using well-defined criteria

Well-defined criteria are essential in archaeoastronomy for
the removal of subjective selection effects which can radically alter
the apparent significance of data. A useful set of selection criteria

Fig.4   Plan of the Barbrook 3 stone circle reduced to
the horizontal plane.

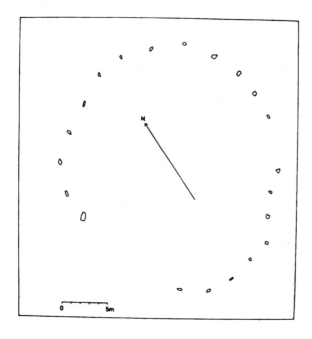

has been suggested by Cooke et al. (1977), and this has been adopted here.
Table 1 lists all the alignments which, according to these criteria, may
be considered as likely astronomical indicators[1]. Fig. 5 shows the horizon
in the direction of one of these. Many potential alignments to outliers
around Barbrook 1 have been rejected because the indicated horizon is less
than 1 km distant.

To evaluate the potential astronomical significance of the
alignments, we have used the statistical test of Freeman and Elmore (1980).
However, this test may be applied only to sets of data that are independent, so the set must not contain, for example, both the alignment from
H to I and the alignment from I to H. We have therefore sorted our data
into four overlapping sets, each of which contains eight independent
alignments. Application of the Freeman and Elmore test to these four sets
yield respective minimum test statistics of -2.2, -2.2, -2.7 and -2.7, at
error angles of 2.2^o and 1.6^o. The astronomical declinations used in the
test were those listed by Freeman and Elmore, namely the maximum and
minimum declinations of the Sun and the Moon, together with the equinoctial
declination of the Sun. The evaluation of the significance of these
results is complicated by the small number of elements in each set of data,
because the statistic is a meaningful indicator of significance only when
the data set contains many elements. We have circumvented this problem
by performing the test on 400 pseudo-random simulations of our data to

Table 1: Indicated Alignments

(ΔAz is the angle in azimuth subtended by the foresight
at the observing position)

	Azimuth		Elevation		Declination		ΔAz
	o	'	o	'	o	'	'
Circle 1-Circle 5	244	58	1	12	-13	39	65
Circle 2-Circle 5	230	29	0	26	-21	59	62
Circle 3-Circle 5	213	32	-0	36	-30	27	26
Circle 5-Circle 1	64	59	-1	07	13	40	62
Circle 5-Circle 2	50	17	-0	47	21	47	59
Circle 5-Circle 3	33	16	0	07	29	53	32
Circle 1-Outlier A	284	43	2	08	10	28	17
Circle 1-Outlier G	232	54	0	48	-20	43	32
Circle 1-Outlier H	229	07	-0	04	-23	07	21
Standing Stone I-Outlier H	141	25	0	35	-27	21	12
Outlier G-Standing Stone I	300	23	2	56	20	04	35
Outlier H-Standing Stone I	321	19	2	43	30	16	36

obtain a frequency distribution of the test statistic. The simulations
show that the likelihoods of obtaining the results of -2.2 and -2.7 by
chance alone are approximately 1 in 10 and 1 in 20 respectively, which
may be considered marginally significant. However, the minimum test
statistic occurs only at large error angles of 1.6° and 2.2°, implying
that any astronomical significance is confined to very rough, perhaps
ritual, alignments.

The Spica Alignment

Thom (1967) noted that outlier A, when viewed from the circle,
indicates the star Spica (α Vir), and from this alignment derived a date
of 2000 BC, which agrees closely with the C14 date obtained by Lewis
(1966).

Our survey confirms the measurements made by Thom. However,
we disagree with his interpretation because of the almost featureless
horizon indicated by A (Fig.6). The azimuth is therefore entirely determined by the line from the circle to the outlier over a distance of about
40m, and depends strongly on the observing position within the circle.

Fig. 5 View from the centre of Barbrook 1 of the alignments
indicated by outliers G and H. In the top left is shown the
path of the setting midwinter Sun in 1800 BC.

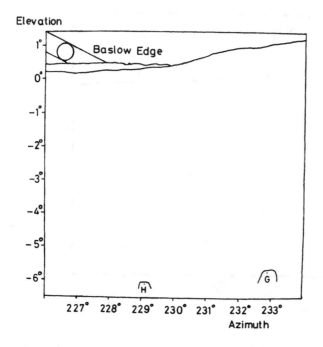

Fig. 6 shows the position of outlier A when viewed from the geometrical centre of the circle. Viewed from elsewhere in the circle, the outlier indicates a declination which may differ by several degrees from that of Spica.

The significance of this alignment therefore rests upon the assumption that the builders of the circle knew the position of its geometrical centre to within a few cm, and hence rests ultimately on Thom's geometrical hypothesis.

Other Alleged Alignments at Barbrook 1

Barnatt (1978) has suggested that 15 astronomical alignments exist at Barbrook 1. One of these (Barnatt's Fig. 51) is from a stone to an unindicated break in the horizon and will not be considered here, as it is not indicated at all on the site. The remaining alignments include 3 to the two sides of a nearby cairn. For conciseness in the discussion below, the two sides of the cairn have been treated as a pair of outliers.

The declinations indicated by these alignments include not only the maximum and minimum rising and setting positions of the Sun and the Moon, but also the rising and setting positions of the Sun on the

Fig. 6   View from the centre of Barbrook 1 of the alignment indicated by outlier A. Shown above the horizon is the path of Spica in 1800 BC

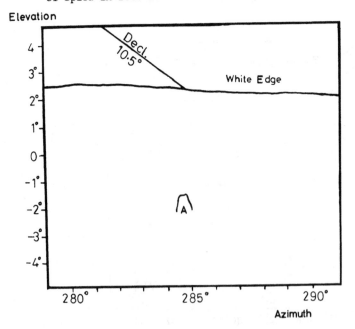

"calendar days" suggested by Thom. There are therefore 13 possible declinations, or 26 azimuths, which should be considered significant in this context.

Barnatt's alignments may be considered in two categories. The first consists of rough ($\pm$ 3^{0}) lines from the centre of the circle to its constituent stones, or from outliers to the circle. There are 21 such lines possible, and 7 of these 'hit' an astronomically significant azimuth. This figure may be compared with the number of hits expected by chance alone, which is $\sim$9.   Clearly these alignments have no statistical significance.

The second category consists of lines between outliers, or from the centre of the circle to the outliers. Taking into account the invisibility of some stones from others, because of topography, there are 87 such lines possible, and 7 of these are astronomically significant. These lines are indicated to an accuracy of $\sim\pm$ $0^{0}.5$, so that the number of 'hits' expected by chance alone is $\sim$6.   These lines also are therefore statistically insignificant.

The conclusion of this section is that, even using the catholic selection criteria of Barnatt, there is no significant evidence at Barbrook for astronomical alignments.

Geometry

Thom (1967) has suggested not only that megalithic sites are astronomical indicators, but also that they were carefully constructed using a knowledge of mathematics and geometry. In particular, he has suggested that Barbrook 1 and 3 are 'Type B' rings. Such a ring shows a degree of flattening which is entirely determined by the geometrical construction.

Using a least squares technique, we fitted in turn to our plans of Barbrook 1 and 3 (Figs. 3 and 4) the shapes suggested by Thom (Types A,B, 1 and 2) and also a circle and an ellipse. The type B shape did indeed give a significantly better fit to Barbrook 1 than any of the other shapes, but at Barbrook 3 the ellipse gave a better (although poor) fit than the Type B ring. There is therefore no evidence for a geometrical construction at Barbrook 3.

The fit of a type B ring to Barbrook 1 is extremely good, although since the largest diameter (17.5 MY) is not an integral number of Megalithic Yards, it does not support Thom's Megalithic Yard hypothesis.

A fit was tried to both the centres of the stones and the inside faces, which in several cases appear to be deliberately flattened; the latter gave a better fit. It should be noted that we have not demonstrated that Barbrook 1 was built according to the Type B construction, but merely that the Type B shape is the closest approximation to the shape of the circle. An alternative hypothesis (Patrick 1978) is that the circles were flattened using an essentially aesthetic rather than geometric construction. Such a hypothesis may be tested adequately only by using an Information Theory technique (Patrick 1978), and then only when a large number of sites have been surveyed. Work in this direction is in progress.

Conclusion

In our survey of the Barbrook sites, and the subsequent data analysis, we have adhered to rigidly defined criteria in order to avoid subjective selection effects. The conclusion of this study is that, according to our criteria, there is no evidence for the Barbrook sites having been designed for any accurate astronomical work, contrary to the suggestions of Thom (1967) and Barnatt (1978). On the other hand, there is marginal evidence of very rough, perhaps ritual, alignments towards significant positions of the Sun and Moon. This conclusion is limited by the small number of alignments at the sites; we hope soon to extend this work to other sites and thereby improve our statistics.

Acknowledgements

We thank Dave Chapel, Alice Few, Tom Muxlow and Cilla Norris for helping with parts of the survey, and Prof. P.R. Freeman for supplying a copy of his program used for the statistical tests. We are particularly grateful to Mr. M. Derrington of the Simon Engineering Laboratories, University of Manchester, for the loan of surveying equipment.

Note

1.   For each alignment, the centre of the foresight and the observing position were estimated at the time of the survey. Consequently the azimuths of two reciprocal alignments may differ slightly from 180° because of the uncertainty in these positions. However, in each case the discrepancy is substantially less than the angle ΔAz.

References

Barnatt, J. (1978). Stone Circles of the Peak. London:Turnstone Books.

Burl, A. (1976). The Stone Circles of the British Isles. New Haven & London: Yale University Press.

Cooke, J.A., Few, R.W., Morgan, J.G. & Ruggles, C.L.N. (1977). Indicated Declinations at the Callanish Megalithic Sites. J.Hist.Astron. 8, 113-33.

Freeman, P.R. & Elmore, W. (1979). A test for the Significance of Astronomical Alignments. J.Hist.Astron.Archaeoastronomy Suppl. 1, 586-596.

Lewis, E. (1966). Some radiocarbon Dates for the Peak District. J.Derbys. Arch. and Nat. Hist.Soc. 86, 115-7.

Patrick, J.D. (1978). An Information Measure Comparitive Analysis of Megalithic Geometries. Ph.D. Thesis, Monash University Dept. of Computer Science.

Radley, J. (1966). A Bronze Age Ring-Work on Totley Moor, and other Bronze Age Ring-Works in the Pennines. Arch.J., 123, 1-26.

Ruggles, C.L.N. & Norris, R.P. (1980). Megalithic Science and some Scottish Site Plans. Antiquity, 54, 40-43.

Thom, A. (1967). Megalithic Sites in Britain. Oxford: Clarendon Press.

OBSERVATIONS AT KINTRAW

T. McCreery
Cardonald College, Glasgow, Scotland

A.J. Hastie
Cardonald College, Glasgow, Scotland

T. Moulds
9 Park Rd, Glasgow, Scotland

Abstract. We have made six visits to the Kintraw site
which has been claimed by Professor Alexander Thom as a
prehistoric astronomical observatory used for the
detection of the midwinter solstice. This claim has been
disputed on several grounds, notably that the foresight is
not visible from the backsight on the ledge overlooking
the menhir. From the results of our observations at the
backsight we contend that this particular dispute is
easily resolved. Other work done by us on the ledge
indicates that it could not have been used as an
observation platform to determine the exact day of the
midwinter solstice.

Of the many megalithic sites claimed by Professor Thom as
remains of prehistoric observatories none has proved more contentious
than the proposed midwinter solstitial site at Kintraw (A. Thom 1971,
pp.37-40). This consists of cairns and a menhir which Thom claims
indicates a horizon foresight, the col between Beinn Shiantaidh and
Beinn a' Chaolais on the island of Jura, 45 km away. Since the view
of this col from the area around the menhir is partially blocked by the
intervening ridge of Dun Arnal, Thom suggests that preliminary
observations were made from a ledge on a hillside to the north east of
the site, and overlooking it. It is the question of the validity of
this claim which has produced such a vigorous debate (MacKie 1976, 1981.
Patrick 1981).

Evidence apparently supporting Thom's suggestion was
furnished by the discovery on the ledge, and at exactly the position
required for a backsight, of two boulders inclined towards each other
so that their inner faces form a V-shaped notch pointing over the
menhir and towards the foresight. Subsequent excavation revealed a
compact layer of stones behind this "boulder notch" (MacKie 1974) and
petrofabric analysis of the layer seemed to indicate that it was
man-made (Bibby 1974, 1982). Other investigators disputed both the

reliability of petrofabric analysis, the use of which in this context is
highly original (McCreery 1980, 1982), and the visibility of the col on
Jura from the boulder notch (Patrick 1981). We attempt here, on the
basis of our own observations from the ledge, to resolve this latter
dispute and to examine more widely the claim that the exact day of the
solstice could have been determined using the ledge as an observing
platform. We offer no comment on the petrofabric analysis since the
viability of the site as a solstitial observatory depends on other, more
important factors. However the presence of a layer of stones in this
area alone may perhaps be explained by the configuration of the slope
above it. Here, there is a shallow couloir which would tend to act as
a funnel channelling all stones moving down this part of the hillside
into the area around the boulder notch and to its left.

      Between September 1980 and October 1981 we visited Kintraw
six times. On each occasion visibility was good enough for the col to
be seen clearly from the hillside above the ledge, and observations
were made with a telescope (x45) from significant positions on the
ledge. The results are presented in Table I.

Table I - Observational Data from the "notch"

Date	Foliage	Eye height above stone layer		Calculated Terrestrial Refraction
		5'0"	5'7"	
27. 9.80	Full	Col obscured	Col obscured	105"
23. 3.81	Sparse	"	Col visible	117"
13. 4.81	Sparse	"	Col obscured	119"
25. 4.81	Sparse	"	Col visible	120"
10. 8.81	Full	"	Col obscured	113"
13.10.81	Sparse	"	Visibility of col varied	

To an observer standing on the stone layer behind the
boulder notch the col was sometimes visible and at other times obscured
by the vegetation on Dun Arnal (Fig. 1). We think that three factors
are responsible for this:-

(A) The observer's height. Changes in eye level of a few
inches are critical.

(B) Unpredictable, but substantial variations in
terrestrial refraction, which changes from hour to hour, along the
length of the alignment from Jura to Kintraw.

(C) The foliage on the clump of trees indicated in Figure 1[1].
In summer this is thick enough to be virtually opaque, but in winter and
early spring, it is sparse enough for it to be just possible to see
through to the skyline beyond. We believe that this factor is less
important than the two mentioned above, for only the tops of the trees
forming the clump rise above the outline of the ridge, as viewed from
the ledge.

Fig. 1 View from the boulder notch. The upper line shows
the maximum observed elevation of the col (optimum conditions)
the lower line shows the minimum elevation. Observer's eye
height 5 ft 7 ins.

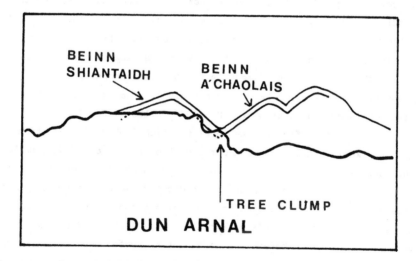

Standing directly behind the boulder notch an observer
whose eye level was 5'0" above the stone layer was never able to see
the col. To an observer whose eye level was 5'7" above the layer the
col was sometimes just visible, rising marginally above the trees on
Dun Arnal, but conditions changed rapidly, presumably because of
variations in terrestrial refraction, on two of our visits. On one of
these occasions, 13th October 1981, at 2.45 p.m. the col appeared
clearer of the ridge than on any previous visit. At 3.30 p.m., after
a squall had passed over Jura, it was obscured by the trees on Dun Arnal,
but ten minutes later was again visible, though barely so. Prior to our
last visit we attempted to correlate our observations with values for
terrestrial refraction calculated using mean values for the refraction
constant (A. Thom 1971, p.31), and meteorological data from the nearest
weather station at Rhuvaal on Islay. These values (Table I) show no
relationship to what was actually observed. This tends to confirm
Moir's suspicion that although Thom used only average values for
terrestrial refraction in his published work (A. Thom 1958), "his actual
raw data ..... was very scattered" (Moir 1981).

These observations enable us to resolve the long standing
dispute between Dr Patrick and Dr MacKie. Patrick has asserted that he
"could not satisfy himself that it was possible to see the base of the
notch using a theodolite with a thirty power magnification," and has
published a photograph to illustrate this (Patrick 1981). Mackie
claims that Patrick's statement is "most incomprehensible" and
"nonsense which Patrick has never explained" (MacKie 1981), and has
also published a photograph (MacKie 1977, p.106). Patrick's
observations, made from an eye level position 5'9" above the stone layer,
are remarkably consistent with our own, from a similar height and his
published statements require no further explanation. MacKie's photograph
shows the col much higher above the ridge than we ever noted in any
observation at the boulder notch. We believe that this may be explained
at least partly by MacKie's personal height. Perhaps also, the
photograph was taken before the ledge was excavated down to the stone
layer. The gain in height resulting from such factors, together with
differing refraction conditions, can explain the discrepancy between
the two published photographs.

This controversy is, however, made meaningless by a further
difficulty. According to Thom, several days before the solstice the
observer on the ledge would have so positioned himself that he saw the
sun flash in the col at the moment of sunset, and would have marked the
position. As the setting position moved southwards the position from
which the observer saw the sun setting in the col moved to the right
(north west) along the ledge till the day of the solstice. On the
following day, as the sunset once more advanced to the north, the
observing position retreated to the left, leaving the extreme right
hand position, subsequently marked by the boulder notch, to identify
the exact day of the solstice. It is evident that this position could
not have been obtained if the positions appropriate to the days
immediately before (and after) the solstice had not also been
established.

At Kintraw the observing position for the day before (and
after) the solstice is 19 ft to the left of the boulder notch. Fig. 2
shows the view from that position.

Fig. 2 View from position 19 ft south east along ledge
from boulder notch showing maximum observed elevation of
Jura hills. Observer's eye height 5 ft 7 ins.

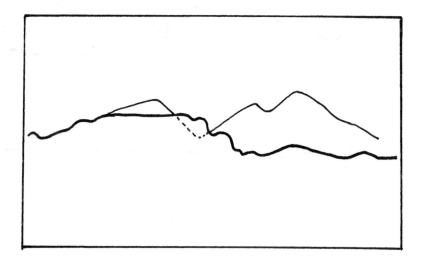

Regardless of conditions the col was always obscured by Dun Arnal itself
and not only by the vegetation on it. This was also true for the
observing positions two and three days before midwinter, at 75 ft and
170 ft to the left respectively. MacKie wrongly states that the
observing position for the day before the solstice is 8.62' to the left
of the boulder notch[2] (MacKie 1976, p.174), but even from this position
the col was invariably obscured, sometimes apparently only by the
vegetation on Dun Arnal, but sometimes by the hill itself (Fig. 3).
It is quite clear, therefore, that the series of observations which
Thom's theory requires could no more have been made from the ledge
than from the field beside the menhir itself.

Fig. 3 View from position 8 ft 6 ins south east along
ledge from boulder notch. The upper line shows the col
obscured by vegetation only (optimum conditions). The
lower line shows the col obscured by Dun Arnal.
Observer's eye height 5 ft 7 ins.

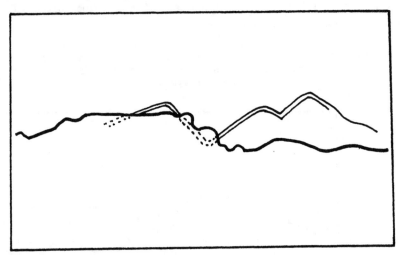

Patrick has suggested that a broad shoulder of the hillside
which he calls a "crest", about 30m above the ledge, might have been a
suitable observing platform, and that one recumbent stone there is
aligned on the Paps of Jura (Patrick 1981. Heggie 1981, p.190).
However if observations of the sun setting in the col were made from
the crest instead of the ledge they would have had to be made from
positions further to the south east, as the altitude of the col is
less from the crest than from the ledge. As the hillside drops away
sharply in this direction there is simply no room for such a series
of positions to be established. There are many large recumbent stones
on this part of the hillside. All lie with their long axes pointing
down the fall line of the slope, which is exactly what is to be
expected if their position is due to natural causes. That they also
point roughly towards the Paps of Jura is coincidental.

Several further objections must be advanced against Thom's
interpretation of the site and to a lesser extent against Patrick's
suggestions. The suggested use of a long alignment means that the
altitude of the foresight is small ($<1^{o}$) and this leads, as we saw
on our visits, to very pronounced changes in terrestrial refraction.
Further changes in the sun's apparent position will be caused by
variations in astronomical refraction and the effect of "graze",
i.e. the rays of light from Jura being bent by Dun Arnal in their
passage to the ledge (A.S. Thom 1981). Finally, the sun will not
generally attain its declination maximum at the instant of setting.
Cumulatively these factors so affect the sun's apparent position that,
even if the visibility of the foresight from the ledge were not in
dispute, it is most improbable that the boulder notch could have been
used as "the permanent observing point from which the day of the
midwinter solstice was regularly checked" (MacKie 1974). Its use
in this context would lead to quite inconsistent measurements of the
length of the year.

We believe that the evidence presented here is not
compatible either with Thom's hypothesis that the ledge could have
been used as an observing platform for work leading up to the
establishment of a permanent backsight at the menhir, or with Patrick's
suggestion that the crest of the hill might have been used in a
similar fashion.

Notes.   1. Due to circumstances beyond our control none of the
photographs we took from the ledge was suitable for reproduction.
The text figures are based both on these photographs and on notes
and sketches made at the time.

            2. MacKie (1974, p.174) has confused the sun's azimuthal
change, which determines the sideways distance to be moved at each
daily observation, with its declination change. At Kintraw, the
azimuthal change is about 28" for the declination change of 12" in
the twenty four hour period before and after the solstice.

References

Bibby J.S. (1974). Petrofabric Analysis. Phil. Trans. R. Soc. London.
        A.276, 191-194.
Bibby J.S. (1982). Forum on Kintraw. Kronos 7 (in press).
Heggie D.C. (1981). Megalithic Science. London: Thames and Hudson.
McCreery T. (1980). The Kintraw Stone Platform. Kronos 5(3), 71-79.
McCreery T. (1982). Forum on Kintraw. Kronos 7 (in press).
MacKie E.W. (1974). Archaeological tests on supposed astronomical
        sites in Scotland. Phil. Trans. R. Soc. London. A.276,
        169-174.
MacKie E.W. (1976) The Glasgow Conference on Ceremonial and Science in
        Prehistoric Britain. Antiquity 50, 136-138.
MacKie E.W. (1977). The Megalith Builders. Oxford: Phaidon.
MacKie E.W. (1981). Wise Men in Antiquity? In Astronomy and Society in
        Britain during the Period 4000-1500 B.C., ed. C.L.N. Ruggles
        & A.W.R. Whittle. B A R British Series 88.
Moir G. (1981). Some archaeological and astronomical objections to
        scientific astronomy in British prehistory. In Ruggles &
        Whittle op. cit.
Patrick J. (1981). A reassessment of the solstitial observatories at
        Kintraw and Ballochroy. In Ruggles & Whittle op. cit.
Thom A. (1971). Megalithic Lunar Observatories. Oxford: Oxford University
        Press.
Thom A.S. (1981). Megalithic Lunar Observatories: an assessment of
        42 lunar alignments. In Ruggles & Whittle op. cit.

Acknowledgements. The authors wish to thank Mr J. Mooney and Mrs M.
        Hastie for their help in preparing the final copy.

DECODING THE CALLANISH COMPLEX - A PROGRESS REPORT

M. R. Ponting
'Olcote', Callanish, Isle of Lewis, PA86 9DZ, Scotland

G. H. Ponting
'Olcote', Callanish, Isle of Lewis, PA86 9DZ, Scotland

Abstract.   Recent research, both archaeological and astronomical, at megalithic sites at Callanish, Carloway and
Bernera is described. Evidence for prehistoric lunar observation at Callanish area sites continues to accumulate;
a moon re-gleam phenomenon occurred on the horizon at the
standstill at most sites so far studied. Difficulties are
experienced in astronomical studies at sites which have not
been investigated archaeologically.

ARCHAEOASTRONOMY AT CALLANISH

The Standing Stones of Callanish on the Isle of Lewis have
been referred to, with some justification, as the 'Stonehenge of the
Hebrides'. They hold a special place in the history of archaeoastronomy.
Quasi-astronomical functions were ascribed to the site by at least
five authors in the 18th and 19th centuries (Toland 1726;   Headrick
1808;   MacCulloch 1824;   Callender 1854;   Kerr 1873).   The archaeo-
astronomers Lewis (1900) and Lockyer (1909) also referred to Callanish.

In 1912, Somerville was the first to complete on-site astronomical research and he identified an alignment to a lunar extreme.
This was the first suggestion that prehistoric man in Britain had
established lunar alignments.   Somerville's paper, and a visit to
Callanish in 1934, were the original inspirations for Thom's detailed
studies on megalithic sites (e.g. Thom 1967; Thom 1971; Thom & Thom
1978).

Hawkins' computer studies of Stonehenge (1963) were followed
by a similar study of Callanish (Hawkins 1965).   Ruggles and his
colleagues' objective methods for determining indicated declinations
were given an initial trial at the sites around Callanish (Cooke et
al. 1977).

DESCRIPTION OF THE SITE

Fig. 1   shows the arrangement of the stones at the main

site, one of about twenty sites around the head of East Loch Roag (Fig.
11). A circle of stones is surrounded by radiating rows of stones.
Two rows form an avenue which runs approximately northwards; two rows
lie roughly east and west, while another runs accurately southwards
from the circle. The largest stone of all, 4.5m tall, stands within
the circle. A chambered cairn also lies within the circle, but recent
excavations (Ashmore 1980; Ashmore 1981) have shown it to be a later
addition to the site.

The layout of 'circle and five stone rows' is an over-
simplification, however. There are three additional stones which do
not fit into this pattern - stones 9, 34 and 35.

THE THREE 'ADDITIONAL' STONES

Hawkins' paper suggested that stone 35 could have been used
as a backsight for three different lunar lines. However, our extensive
documentary researches (Ponting & Ponting, in press) indicated that
stone 35 had been re-erected last century, following its discovery
prone in the peat. The recent excavations confirmed our conclusions -
the stone proved to have been set in concrete. Its present position
was probably chosen about 1860. Excavation showed that the prehistoric
socket hole was not below the present position of the stone, nor could
the socket be identified in the surrounding area. Hawkins' astronomical
conclusions were based on an invalid position of the stone.

Stones 9 and 34 form an alignment across the circle. Stone
34 is set out of line with, and at an angle to, the other stones in
the east side of the avenue. It is a massive slab, but from stone
9 it is seen end-on and acts as a slender pointer for the horizon above.
Somerville's claim that this indicated the northern extreme moonrise
at maj. standstill has been confirmed by our own survey (Ponting &
Ponting 1981a). Excavation did not bear out the idea (Burl 1976) that
stone 9 is the remnant of an outer circle, and the astronomical theory
is the only one proposing an explanation for the positions of stones
9 and 34.

The new information on the Callanish site, resulting from
Ashmore's excavations, highlights the dilemma (c f. Heggie 1981) for
the archaeoastronomer using sites or stones of uncertain archaeological
status. Following the documentary search and the excavations, we feel
confident that the other stones at the main Callanish site have stood

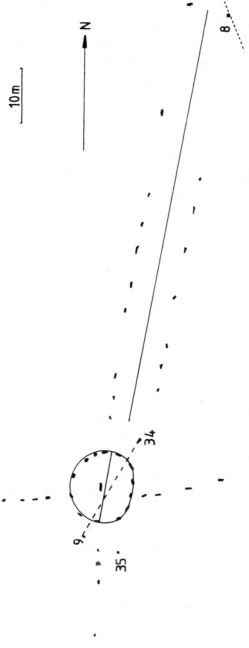

Fig. 1. Plan of Callanish I showing possible astronomical alignments. Solid lines show the type A flattened circle with its long axis, and the central line of the avenue; these are parallel and the avenue line may be a ritual indication of the $-(\epsilon+i)$ moonset. The long-dashed line shows the alignment 9–34, an indication of $+(\epsilon+i)$ moonrise; the set of the same moon is indicated by the attitude of the flat face of stone 8 (shown by the short-dashed line). Stone numbering after Somerville (1912).

Fig. 2. Path of the moon at $-(\epsilon+i+\Delta)$ in 1550 B.C., as seen from the centre of the north end of the avenue. The avenue alignment indicates a symbolic set of the moon into the stones of the circle. Figures on the horizontal scale are degrees of azimuth; those on the vertical scale are degrees of apparent altitude.

in their prehistoric positions to the present day, and that few stones
are missing. Thom & Thom (1978) are mistaken in their statement concerning the re-erection of stones at Callanish.   Worsaae's sketch (1846)
unequivocably shows the complete circle embedded in peat.   The peat
protected the stones and, indeed, hid the cairn from view till it was
'excavated' in 1857 (Innes 1858).

HORIZON SURVEYS

In a previous paper (Ponting & Ponting 1981a), we outlined
the aims and methods of our archaeoastronomical research at Callanish.
To obtain the initial results quoted in that paper, we had concentrated
on the southern extreme of the major standstill as viewed from a variety
of sites.

Our ultimate aim is to complete a 360° horizon survey from
each of the Callanish sites.   In this way we shall accumulate a large
corpus of data, of a rather different type from that collected by other
researchers.   It will be suitable for statistical analysis; for example,
the significance of any apparent astronomical indications at a site
could be checked by the computerised random rotation of the site,
relative to its horizon.

SOUTHERN EXTREME OF THE MAJOR STANDSTILL

At the latitude of the Callanish sites (58° 12' N) the
southern extreme of the major standstill is a spectacular event. In prehistoric times, the moon's centre never rose more than 2° 10' above
a level horizon. Its brief, low passage occurred within the azimuth
range 160° to 200°. Within these azimuth limits, as seen from Callanish,
there are two prominent ranges of hills: the hills of Pairc in SE
Lewis, known locally as 'the Sleeping Beauty' from their similarity
to a supine female figure (Fig. 4);   and the hills of North Harris,
dominated by Clisham (799m) and the deep V-notch of Glen Langadale
(Fig. 5).

At its southernmost extreme, the moon rose out of the
Sleeping Beauty and set into the Harris hills.   This is true for an
observer anywhere within an area which we have called 'the Callanish
Diamond' - and all the Callanish sites are within this Diamond (Fig.
3, in which some dubious sites within the Diamond have been omitted).

What is more, in almost all cases so far studied, after

Fig. 3. The hypothetical 'Callanish Diamond'. For explanation see text.

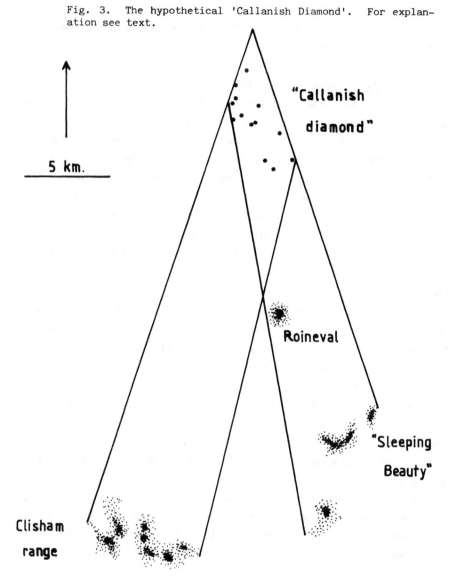

"Callanish

diamond"

5 km.

Roineval

"Sleeping

Beauty"

Clisham

range

Fig. 4. The range of hills known to Callanish people as
the 'Sleeping Beauty'. (Mor Mhonadh = knees, Sidhean an
Airgid = face).

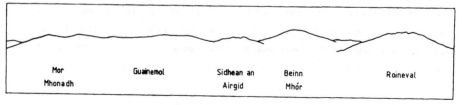

| Mor | Guainemol | Sidhean an | Beinn | Roineval |
| Mhonadh | | Airgid | Mhór | |

setting into the Harris hills, the moon re-appeared ('re-gleamed') in the V-notch, in such a way that perturbations could have been distinguished (e.g. Fig. 5).

THE CALLANISH MINOR SITES

We have detailed (Ponting & Ponting 1981a) 20 other known and putative megalithic sites in the Callanish area (see Fig. 11). While the more obvious of these 'minor sites' were known to antiquarian visitors last century, none of them has been investigated archaeologically, so we do not know if they are contemporary with the main site and with the various single megaliths and groups of megaliths.

However, many of the sites are intervisible and we intend to continue investigating possible astronomical relationships between the sites. From our studies so far, several of the single megaliths and two complex but non-circular sites could have been used for lunar observation and recording.

Four of the minor sites are stone rings, and together with the ring at the centre of Callanish I, they have geometrical and astronomical similarities. The rings at Cnoc Fillibhir Bheag (Callanish III), Na Dromannan (Callanish X) and Callanish I all match closely the shape which Thom (1967) has called a type A flattened circle. The rings at Cnoc Ceann a' Gharaidh (Callanish II) and Ceann Hulavig (Callanish IV) are both ellipses - of different sizes, but identical proportions. We believe that there can never be absolute certainty when fitting a geometrical shape to the plan of a site, so there is always some play in the direction of the axis of a flattened circle or an ellipse. Nevertheless, we consider that at four of the rings mentioned, the long axis indicates (even if only at a symbolic level) a horizon position of the moon at the major standstill. The fifth site (Ceann Hulavig) has a long axis which indicates midsummer sunset over Callanish I.

Cleiter (Callanish VIII) and Airigh Mhaoldonuich (Callanish XV) are both on the island of Great Bernera in Loch Roag. Although not in the Callanish Diamond, both of these sites also appear to be related to the southern extreme path of the moon. At the fallen monolith, Callanish XV, the re-gleam occurred as at many of the Callanish sites, though over a different range of hills.

Fig. 5.   -(ε +i- Δ) moonset as seen from Druim na h'Aon
Chloich (site XVII) with re-gleam over site IV in the V-notch.

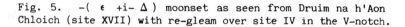
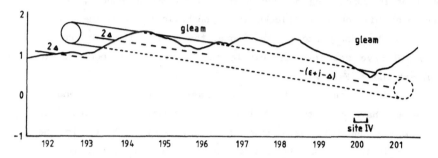

Fig. 6.   +(ε + i+ Δ) moonset as seen from stone 8 at the
main site.   Thick line below notch represents probable prehistoric ground level.

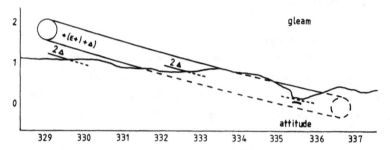

Fig. 7. +(ε+i) moonset as indicated by the NE face of the
erect stone at Clach an Tursa, Carloway.

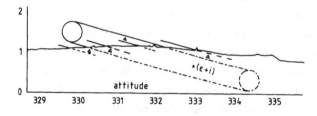

Fig. 8.   Path of the moon at -(ε+i+ Δ) as it transits on Bheinn
Iobheir, as seen from Clach an Tursa.

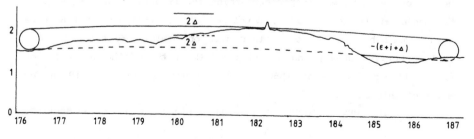

MORE ASTRONOMY AT THE MAIN SITE

We have shown (Ponting & Ponting 1981a) that the Thom and Hawkins theories concerning moonset into Clisham are untenable, due to an intervening hillock which blocks the hypothetical sight-line. But, as viewed from the north end of the avenue, the moon at its southernmost extreme would have set into the circle itself, surely a good example of ritual astronomy (Fig. 2).

In fact, our results so far suggest that the north end of the avenue may have been the most important position at the main site for astronomical observations, rather than any point near or in the circle. For instance, the attitude of stone 8, which is not aligned with the rest of its row, indicates a notch where the setting moon re-gleamed (Fig. 6) - the same northern extreme winter moon whose rise was indicated by the line 9-34.

CLACH AN TURSA, CARLOWAY

10km north of Callanish is a less well-known setting of three stones, known as Clach an Tursa (N.G.R. NB 204430). One stone, 2.1m tall by 1.1m by 0.6m, stands erect, but the other two lie prone and broken on either side, 5.1m and 4.7m long respectively. Martin (1716, 8) definitely stated that three stones were erect at Carloway, each about 12 feet tall. The short description by R.C.A.M.S. (1928, 26-7, no. 89) agrees essentially with the state of the site as it is today (see Figs. 9 & 10). It is impossible to be certain, without excavation, whether the three stones formerly stood in line, though this seems likely from their relative positioning. It is just conceivable that they formed part of a circle, but there is no visible evidence for this.

The site is situated on the croftland of Upper Carloway, within a basin of hills at the head of Loch Carloway, an inlet of Loch Roag. Only 9° of the 360° horizon is formed by hills more than 4km away.

Our intention in studying the Carloway site was to check whether our working hypothesis concerning the importance of the re-gleam phenomenon at the standstills, developed at the Callanish and Bernera sites, could be applied to another site on Lewis. We carried out a complete horizon survey and superimposed the paths of the sun and moon for significant events in 1550 B.C. (see Ponting & Ponting 1981a for choice of this date).

The erect stone is roughly rectangular in cross-section,

Fig. 9.   Ground plan of Clach an Tursa, Carloway.   This survey of the site, by theodolite and tape, is believed to be the first published (Ponting & Ponting 1981b).   The attitude of the erect stone (dashed arrows) is discussed in the text.   The two fallen and broken stones are shown in outline.

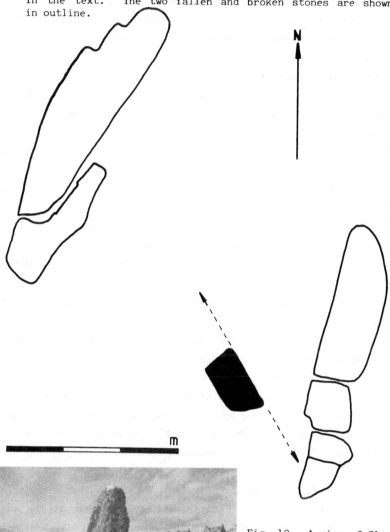

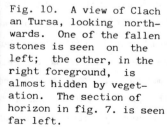

Fig. 10.   A view of Clach an Tursa, looking north-wards.   One of the fallen stones is seen on the left;   the other, in the right foreground,   is almost hidden by veget-ation.   The section of horizon in fig. 7. is seen far left.

and we ignored the attitudes of its shorter faces. As a flattened slab
is often aligned with the stone row in which it stands, we have assumed
that the attitude of the stone remaining erect represents the original
alignment of the supposed row. Of the two long faces, the SW face is
slightly convex and no attitude can be determined with certainty. We
therefore concentrated our attention on the NE face, which is flat;
although it is not a perfect plane and it has a luxuriant growth of
lichens, it is possible to determine its attitude.

Sighting in the northerly direction along the NE face, a
hillock is indicated with horizon boulders on it, 1.3km away. It is
clear from our profile (Fig. 7) that the hillock lies within the lunar
band for the setting moon at $+(\epsilon + i \pm \Delta)$. In fact, two of the boulders
– at azimuths $330°20'$ and $331°55'$ – closely match the declinations of
the upper limb and the lower limb of the setting moon, in its average
position between the two perturbations.

We accept that these boulders require further investigation,
especially as there are several boulders and slabs of rock in this one
area, which have voids beneath and so could be fallen megaliths. It
is also interesting that an elderly inhabitant remembers stories of a
second 'Tursachan' (group of standing stones) in Carloway, in addition
to Clach an Tursa.

Sighting along the NE face of the stone in a southerly
direction, a small notch is indicated at a junction between a relatively
smooth section of horizon and a more hummocky section. This notch has
an azimuth of $150°35'$ and an altitude of $55'$ which gives a declination
which is too far south for any solar events, but not far enough south
to fall in the lunar band for $-(\epsilon + i \pm \Delta)$. At moonrise, this lunar band
extends across the Carloway horizon between azimuths $163°29'$ and $168°$
$50'$.

As Carloway is five minutes of latitude north of Callanish,
the moon's extreme southern path is even lower here. To the south of
the site, Bheinn Iobheir is 1.1km away. The name translates as Ivor's
Hill; it stands 77m high and is topped by Jubilee Cairn. We have been
told locally that it was built by villagers in 1897 in a position which
has been the focus of celebrations, often with bonfires, since time
immemorial. The moon's transit occurred very close to the position of
the present cairn, and at $-(\epsilon + i + \Delta)$ only a small portion of the
equinoctial half moon would have remained visible above the horizon at

transit (Fig. 8).

        In fact, the moon would have completely disappeared, were
the difference in site and horizon ground levels only 0.5m greater than
it is today. Limited probing near the erect stone suggests that the
prehistoric ground level was about 0.3m below the present surface,
although excavation would be needed to confirm this. The hilltop around
the cairn is bare rock, which may well have had a soil covering in prehistoric times. The effect of both these factors is to raise the horizon
relative to the moon's path, altering the relationship shown on the
profile. Therefore, the disappearance of the moon at transit is probable
and this would have been followed by a re-appearance or 're-gleam' before
it finally set.

        Northern extreme moonset appears to have been indicated
at Clach an Tursa, and it is also possible that the re-gleam phenomenon
behind Bheinn Iobheir may have been important in the choice of the site.
Thom has claimed that a well-placed lunar observatory can use two or
more foresights in different directions. Although no part of the moon's
path at $-(\epsilon + i)$ is indicated today, it is clear that a full understanding
of the Carloway site depends upon a knowledge of the positions of the
socket-holes of the fallen stones. Excavation would be required to
determine these positions; the hypothesis that Clach an Tursa may have
been a lunar observatory would inevitably be strengthened or weakened
by this knowledge.

FUTURE WORK

        The recent discovery (Ponting & Ponting 1981c) of a
previously unrecorded stone circle, with at least 18 megaliths and a
diameter of 41m at Achmore, 11km from Callanish, underlines how much
archaeological work remains to be done on the Isle of Lewis.
Unfortunately, the Islands Council does not employ an archaeologist
and there is no local structure for survey, recording and rescue.

        While continuing with our astronomical studies at the sites
around Callanish, we shall undertake archaeological work within our
competence, as it arises. We have several years' work ahead of
us before we can consider writing the book which we might be tempted
to call 'Callanish Decoded'.

Fig. 11.   The head of East Loch Roag, showing the positions
of the Callanish area sites.   Carloway is shown in the
location map.   All site names, grid references, etc., are
tabulated in Ponting & Ponting 1981a .

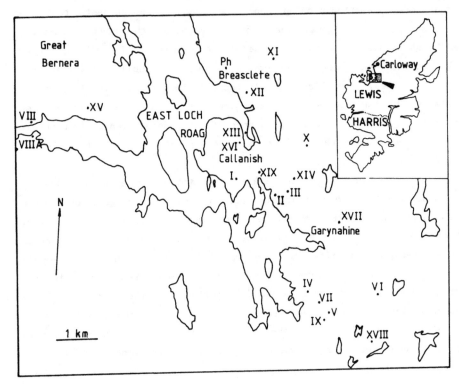

ACKNOWLEDGEMENTS

Thanks are due to the following: Patrick Ashmore of the
Ancient Monuments Branch of the Scottish Development Department for
permission to quote from the interim results of his excavation at
Callanish;   Linda Louden and J. N. G. Ritchie of the R.C.A.M.S. for
assistance in the preparation of this paper for publication;   Comhairle
nan Eilean (Western Isles Islands Council) for financial assistance
with our attendance at the Oxford Symposium;   and the crofters of
Callanish, Carloway and Bernera for permission to work on their land.

REFERENCES

Ashmore, P. (1980). Callanish 1980 Interim Report. Edinburgh: Scottish
         Development Department.
Ashmore, P. (1981). Callanish 1981 Interim Report. Edinburgh: Scottish
         Development Department.
Burl, A. (1976).   The Stone Circles of the British Isles.   New Haven
         and London: Yale University Press.
Callender, H. (1854). Notice of the Stone Circle at Callernish. Proc.
         Soc. Antiq. Scot., 2, 380-4.
Cooke, J. A.,Few, R. W., Morgan, J. G., and Ruggles, C. L. N. (1977)
         Indicated Declinations at the Callanish megalithic sites.
         J. Hist. Ast., 8, 113-33.
Hawkins, G. S. (1963). Stonehenge Decoded. Nature, 200, 306-8.
Hawkins, G. S. (1965). Callanish, a Scottish Stonehenge. Science, 147,
         127-30.
Headrick, J. (1808). Editorial notes in G. Barry, History of the Orkney
         Isles. (2nd edition) London.
Heggie, D. C. (1981). Highlights and Problems of Megalithic Astronomy.
         J. Hist. Astr., 12;   Archaeoastronomy supp., 3, S17-S37.
Innes, C. (1858). Notice of the Stone circle of Callernish in the Lewis,
         and of a chamber under the circle recently excavated. Proc.
         Soc. Antiq. Scot., 3, 110-12.
Kerr, J. (1873).   The cruciform sun temple at Callernish. The Builder,
         12 July 1873.
Lewis, A. L. (1900). The stone circles of Scotland. J. Anthrop. Inst.,
         30, 56-72.
Lockyer, N. (1909).   Stonehenge and other British Stone Monuments
         Astronomically Considered. (2nd edition) London: Macmillan.
Martin, M. (1716).   A Description of the Western Islands of Scotland.
         London: Bell. Reprinted 1970, Edinburgh: Mercat Press.
MacCulloch, J. (1824).   The Highlands and Western Islands of Scotland.
         London.
Ponting, M. & Ponting, G. (1981a).   Decoding the Callanish Complex –
         some initial results in Astronomy and Society in Britain
         during the period 4000-1500 B.C., ed C. L. N. Ruggles & A.
         W.R.   Whittle,   63-110. Oxford: British Archaeological
         Reports,British Series, no. 88.
Ponting, G. & Ponting, M. (1981b). Clach an Tursa, Carloway. Callanish:
         privately published.
Ponting, M. & Ponting, G. (1981c). Achmore Stone Circle. Callanish:
         privately published.
Ponting, G. & Ponting, M. (In press). The Stonehenge of the Hebrides.
         Findhorn: Thule Press.
R.C.A.M.S. (1928).   Royal Commission on the Ancient and Historical
         Monuments of Scotland. Inventory of Monuments ..... in the
         Outer Hebrides, Skye and the Small Isles. Edinburgh: H.M.S.O
Somerville, B. (1912). Prehistoric monuments in the Outer Hebrides and
         their astronomical significance. J. R. Anthrop. Inst., 42,
         15-52.
Thom, A. (1967). Megalithic sites in Britain. Oxford: Oxford University
         Press.
Thom, A. (1971).   Megalithic lunar observatories.   Oxford: Oxford
         University Press.
Thom, A., & Thom, A. S. (1978).   Megalithic Remains in Britain and
         Brittany. Oxford: Clarendon Press.
Toland, J. (1726).   A History of the Druids. (New edn. London, 1814).
Worsaae, J. J. A. (1846).   Manuscript field notebooks (Unnumbered).
         Copenhagen: Nationalmuseet.

ASTRONOMY AND STONE ALIGNMENTS IN S.W. IRELAND

A. Lynch
National Monuments Branch, Office of Public Works,
51 St Stephen's Green, Dublin 2

Abstract.   The Bronze Age stone alignments of S.W. Ireland were surveyed in detail to provide data with which to test hypotheses concerning their astronomical, metrological and morphological attributes. The astronomical tests are described in this paper. The hypothesis examined was that the centres of the stones of an alignment define a line which is orientated on an event of astronomical significance. The events considered as significant were the lunar standstill positions, the solstices and equinox. The alignments were treated as open-ended structures and their azimuths in both directions were considered. Two tests were carried out. In one, the allowable azimuth error was calculated individually for each site, and in the other, an average value was used. The probability level at which the hypothesis operates was calculated from a version of Bernoulli's Theorem and the results from both tests allow us to accept the hypothesis.

INTRODUCTION

This paper is based on the results of a study carried out in the period 1973 to 1976 in which a detailed survey was made of the stone alignments of S.W. Ireland and hypotheses concerning their astronomical, metrological and morphological attributes were tested (Ni Loingsigh 1976). Only the astronomical results are considered here.

The definition of a stone alignment, as accepted for this work, is three or more standing stones, intervisible and in a straight line. There are two major concentrations of such sites in Ireland, one in the southwest (i.e. counties Cork and Kerry) and the other in Northern Ireland (counties Derry, Tyrone and Fermanagh in particular). The total number of alignments in the southwestern group is 39 and 37 of these were surveyed to provide data for this study (Table 1).

Before dealing with their astronomical aspects, a brief assessment of the archaeological background of the stone alignments will be made. The dating of these sites on archaeological grounds remains problematical. Few of the Irish sites have been excavated and of these

Table 1. Site data. All measurements are degrees decimal except Latitude
indicated are those

Site	Lat.N.	Long.W.	Azimuth	Horizon Altitude	Declination	Tole-rance	Target
Newcastle	51°58'	8°37'	230.16	-0.44	-23.74	0.10	None
			50.16	1.00	24.12	0.10	None
Castlenalacht	51°48'	8°44'	238.06	-0.02	-19.18	1.02	S.Lunar Min.
			58.06	-0.37	18.80	1.02	None
Gurranes	51°32'	9°11'	243.80	0.03	-16.41	2.42	None
			63.80	-0.56	15.49	2.42	None
Cabragh (A)	51°58'	9°03'	228.74	0.03	-24.36	2.42	Winter Solstice
			48.74	9.44	32.02	2.42	None
Cabragh (B)	51°58'	9°03'	232.19	0.03	-22.30	0.48	None
			52.19	0.03	22.23	0.48	None
Cullenagh	51°44'	9°13'	210.62	0.33	-32.16	0.02	None
			30.62	1.43	33.54	0.02	None
Dromdrasdil	51°44'	9°11'	233.96	0.20	-21.75	1.07	None
			53.96	4.09	24.82	1.07	None
Tullig	52°02'	8°59'	244.22	1.40	-14.53	0.15	None
			64.22	4.07	18.84	0.15	None
Knockawaddra	51°39'	8°59'	209.04	0.83	-32.11	0.57	None
			2.04	-0.56	32.34	0.57	None
Farranahineeny	51°48'	9°09'	222.37	0.30	-27.12	0.31	None
			42.37	0.48	27.63	0.31	None
Maughanasilly	51°46'	9°23'	219.58	0.94	-28.01	0.96	None
			39.58	1.34	29.77	0.96	None
Monavaddra	51°49'	9°11'	236.74	-0.51	-20.68	0.57	None
			56.74	3.74	22.94	0.57	None
Ardrah	51°44'	9°21'	237.13	1.76	-18.28	3.91	S. Lunar Min.
			57.13	-0.29	19.41	3.91	N.Lunar Min.
Kilcaskan	51°42'	9°43'	206.02	14.88	-19.71	0.74	S. Lunar Min.
			26.02	-0.32	33.56	0.74	None
Maulinward	51°39'	9°28'	216.89	1.90	-28.43	1.19	None
			36.89	4.28	33.62	1.19	None
Leitry Lower	51°42	9°50'	213.54	-0.34	-31.03	0.69	S. Lunar Max
			33.54	2.73	33.73	0.69	None
Coolacoosane	51°58'	8°59'	203.90	1.95	-32.71	1.89	None
			23.90	0.14	34.43	1.89	None

and Longitude. The horizon altitude is the true altitude. The targets obtained in Test 1.

Site	Lat.N.	Long.W.	Azimuth	Horizon Altitude	Declination	Tolerance	Target
Reananerree	51°54'	9°09'	213.10	0.18	-31.03	0.47	None
			33.10	0.37	31.47	0.47	None
Cloonshear Beg	51°51'	9°04'	232.54	0.01	-22.38	0.82	None
			52.54	1.35	23.23	0.82	None
Dromcarra Nth.	51°51'	9°03'	222.72	-0.41	-27.72	0.43	None
			42.72	2.08	28.84	0.43	None
Slievereagh	51°58'	9°11'	242.58	7.09	-10.98	0.36	None
			62.58	5.23	20.76	0.36	None
Glantane East	52°00'	9°03'	242.10	4.93	-12.76	0.76	None
			62.10	2.55	18.84	0.76	None
Carrigagulla	52°00'	8°55'	207.02	1.40	-31.97	0.07	None
			27.02	-2.84	30.60	0.07	None
Rossnakilla	51°50'	8°59'	223.21	1.94	-25.19	0.22	None
			43.21	1.16	27.80	0.22	None
Beenalaght	52°02'	8°45'	220.56	2.36	-26.04	0.59	None
			40.56	0.70	28.50	0.59	None
Garrane	52°04'	8°46'	215.03	-0.32	-30.56	1.21	None
			35.03	8.40	37.83	1.21	None
Gneeves	52°06'	8°46'	273.70	2.15	3.51	0.77	None
			93.70	1.49	-1.09	0.77	None
Dromteewakeen	51°58'	9°48'	233.49	6.93	-15.88	0.63	None
			53.49	2.96	24.01	0.63	Summer Solstice
Cashelkeelty	51°45'	9°49'	262.25	2.24	-3.21	0.02	None
			82.25	3.94	7.90	0.02	None
Kildreelig	51°48'	10°19'	243.24	0.68	-15.78	0.64	None
			63.24	14.92	28.14	0.64	N.Lunar Max
Eightercua	51°49'	10°09'	227.29	0.97	-24.13	0.62	Winter Solstice
			47.29	0.41	25.16	0.62	None
Doory	51°52'	10°07'	216.26	0.03	-29.98	0.76	S.Lunar Max.
			36.26	3.04	32.62	0.76	None
Dromkeare	51°49'	10°10'	201.24	2.49	-32.88	0.68	None
			21.24	5.41	40.38	0.68	None
Cloonsharragh	52°14'	10°11'	241.19	0.03	-17.28	0.66	None
			61.19	7.01	22.93	0.66	None
Ardamore	52°08'	10°01'	225.90	1.98	-24.02	0.24	Wint.Sol.
			45.90	0.62	25.84	0.24	None
Beal Middle	52°32'	9°40'	215.64	1.32	-28.71	0.62	None
			35.64	0.20	29.83	0.62	None
Dromatouk	51°52'	9°32'	227.21	0.94	-24.12	0.14	None
			47.21	2.63	27.09	0.14	None

only two produced any sort of dating evidence. The construction of the
stone circle/alignment complex at Beaghmore, Co. Tyrone has been bracketed
by the dates 1535±55 bc and 775±55 bc (Pilcher 1969, 69) while the
construction date of the recumbent-stone circle at Cashelkeelty, Co. Kerry,
lies between 970±60 bc and 715±50 bc (Lynch 1981, 66)[1]. The stone
alignment at Cashelkeelty was constructed shortly before the adjacent
circle, which would place it c. 1000 bc. On present evidence the stone
alignments of Ireland may be regarded as belonging to the second half of
the second millenium bc (i.e. the later Bronze Age). This period in
S.W. Ireland was one of intense activity. Widespread clearance of
woodland was taking place, agriculture was well established and the copper
ore deposits of the region were being exploited. Recent palynological
studies have shown that several of the alignments were built on abandoned
farmland which was reverting to heath and that the immediate environment
of the alignments was open scrubland (Lynch 1981, 125).

ASTRONOMICAL ASPECTS

Hypothesis

Prompted by the work of Alexander Thom on the British megaliths
(e.g. Thom 1954, 1955, 1964, 1966, 1967 and 1971) and that of John Barber
on the recumbent-stone circles of S.W. Ireland (Barber 1973), it was
decided to test the hypothesis that the centres of the stones of an
alignment define a line which is orientated on an event of astronomical
significance.

Definitions and Methodology

The alignments were surveyed using a Kern theodolite. The plans
were drawn on the horizontal plane at ground level and each survey point
was reduced by computer to (x,y) coordinates. By the simple process of
averaging the x and y values, the centre of mass of a lamina, the shape of
the base of the stone, was derived. Since the stones are quite regular
this would never have been far from the line of operation of the centre of
mass of the stone.

The term orientation implies a single and specific line
pointing in some direction on the earth's surface. A least squares
adjustment was used to find the line of best fit to the centre of the
stones and the horizontal distance errors (H.D.E.) were calculated for
each site (i.e. the distances of the centre of the stones from the line of
best fit). Before continuing with the test, the criterion that the H.D.E.

should not be significantly different from zero was imposed, if the lines
of orientation were to be accepted. A Student's t test was applied to the
H.D.E. of all the sites and the resultant t value was 0.6381 which is not
significantly different from zero.

The azimuth of the line of orientation was derived by
constructing a magnetic closure of four points enclosing the alignment
at the time of survey. The error of the magnetic closure gave the error
due to local magnetic anomalies. The closure was so laid out that one
diagonal was always roughly parallel with the alignment. The magnetic
bearing of this diagonal was corrected for the secular magnetic variations
(annual, seasonal & diurnal) obtained from data kept by Valentia
Observatory, Co. Kerry. This resulted in getting the True Bearing of the
diagonal and the azimuth of the line of orientation was then calculated
as part of the computer programme (courtesy of Dr. I.O. Angell, Royal
Holloway College, Egham, Surrey) by calculating the angle between the
line of best fit and the surveyed diagonal. Alignments do not, of course,
point in one direction only and their azimuths in both directions were
considered (Fig. 1).

The events considered as astronomically significant were the
turning points of the solar cycle and its midpoint (i.e. Summer and Winter
Solstices and the Equinox) and the extreme or standstill positions of the
moon (Table 2).

Table 2. Astronomically significant events.

Event	Declination	Event	Declination
Summer Solstice	$+23°.91$	S. Limit of Major Lunar Standstill	$-29°.95$
Winter Solstice	$-23°.91$	N. Limit of Major Lunar Standstill	$+28°.17$
Equinox	$0°$	S. Limit of Minor Lunar Standstill	$-19°.58$
		N. Limit of Minor Lunar Standstill	$+17°.94$

The solar declinations take into account the change in the obliquity of the ecliptic (ϵ) over the years. Thom (1971,15), using De Sitter's formula to determine ϵ, gives a value of $23^{\circ}.91$ for ϵ at 1800 BC and this is the value used here. The effects of parallax have been taken into account for the lunar declinations.

The first step in calculating the probability level at which the astronomical hypothesis operates was to calculate the declination, east and west, for the line of best fit at each site. This was done in two different ways. In Test 1, the accuracy to which the orientation of each monument could be defined was calculated by getting the Arctan of the average H.D.E. for the site, divided by the length of the alignment. Fig.1 Alignment orientations.

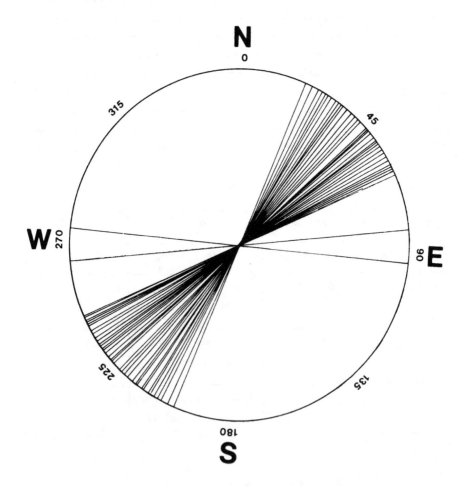

The resulting quantity (the uncertainty angle or tolerance, as presented
in Table 1) was first added to the azimuth and then subtracted from it
and in each case the declination was calculated, to give a small angular
spread within which, if the monument was astronomically orientated, the
declination of an astronomically significant target, as defined above,
should be found. A similar approach was adopted in Test 2 except that an
average uncertainty angle (1.791°) was calculated for the group as a
whole.

The probability level was calculated from a version of
Bernoulli's Theorem, which gives the chances of scoring (x) hits out of
(n) shots at targets which occupy a proportion (p) of the target area:

$$\frac{n!}{x!\,(n-x)!}\; p^{x}(1-p)^{n-x}$$

To calculate (p) we must turn to the situation as it confronted the
alignment builders. They were free to orientate the alignments in any
direction in the horizontal plane, therefore 360° is the total possible
target area. As shown above, there were seven possible solar/lunar targets
on the western and eastern horizons which gives a total of fourteen.
Finally, the azimuth range considered at each site extends over 2(uncertainty angle).
Therefore

$$P = \frac{14 \times 2(\text{uncertainty angle})}{360}$$

Results.

In Test 1, ten sites had significant astronomical orientations
(Table 1) and of these, nine were astronomical in one direction only
(seven western and two eastern). The total of eleven 'hits' included
seven lunar standstill declinations and four solsticial declinations.
The probabality level at which the hypothesis operates, using this
particular analysis, is 0.0024, which allows us to accept the hypothesis.[2]

In Test 2, of the thirty seven sites examined, twenty three
had significant astronomical orientations and of these, twenty one were
significantly orientated in one direction only (thirteen eastern and
eight western). The total of twenty five 'hits' included thirteen lunar
standstill declinations, eleven solsticial declinations and one equinoctial
declination (Gneeves). This result, at a probability level of 0.0000089,

is more significant statistically than that obtained in Test 1.

COMMENTS.

It may be reasonably concluded from the above results that the astronomically significant orientations of the stone alignments are not accidental. Looking at the results of both tests there seems a slight preference for lunar orientations but, whether solar or lunar, the events indicated are the more obvious from the observer's point of view and do not imply a highly scientific group of alignment-builders.

The spirit in which the alignments were orientated remains obscure. The diverse events indicated within such a homogenous group of sites would tend to rule out a practical explanation such as regulating a calendar. One may, however, speculate on the possible symbolic or ritualistic function of the alignments. The excavations of the stone circle/alignment complex at Cashelkeelty, Co. Kerry, indicated a close association with burial rites and it may well be that the astronomical alignments served some symbolic function in relation to burial ritual. Future excavation will hopefully increase our knowledge of the social and economic environment of these alignment-builders and this in turn may help in our evaluation of their more mystical attributes.

NOTES

1. The notation bc/ad is used for uncalibrated radiocarbon dates.

2. Even though the uncertainty angle was calculated individually for each site, and this was the value used to determine the declinations and resultant 'hits', it was the average uncertainty angle of all sites (1.791^{o}) which was used in calculation of p and the probability level.

REFERENCES

Barber, J. (1973). The orientations of the recumbent-stone circles of
          the south-west of Ireland. Journal of the Kerry Archaeological
          and Historical Society, *6*, 26-39.

Ni Loingsigh, A. (A. Lynch) (1976). The Stone Alignments of Cork and
          Kerry. M.A. Thesis, National University of Ireland, University
          College, Cork.

Lynch, A. (1981). Man and Environment in S.W. Ireland. British
          Archaeological Reports, British Series 85.

Pilcher, J.R. (1969). Archaeology, palaeoecology and C14 dating of the
          Beaghmore stone circle site. Ulster Journal of Archaeology,
          32, 73-91.

Thom, A. (1954). The solar observatories of megalithic man. Journal
          of the British Astronomical Association, *6*, 396-404.

Thom, A. (1955). A statistical examination of the megalithic sites in
          Britain. Journal of the Royal Statistical Society, *118A*,
          275-295.

Thom, A. (1964). Observatories in ancient Britain. New Scientist, *398*,
          17-19.

Thom, A. (1966). Megalithic astronomy : indications in standing stones.
          Vistas in Astronomy, *7*, 1-57.

Thom, A. (1967). Megalithic Sites in Britain. Oxford : Oxford University
          Press.

Thom, A. (1971). Megalithic Lunar Observatories. Oxford : Oxford
          University Press.

STONE RINGS OF NORTHERN POLAND

ROBERT M. SADOWSKI

ul. Międzynarodowa 33 m. 24, 03-962 Warszawa, Poland

MARIUSZ S. ZIÓŁKOWSKI

KAROL PIASECKI

Zakład Antropologii Historycznej UW, Krakowskie
Przedmieście 1, 00-927 Warszawa, Poland

Abstract. This paper is a preliminary account of field research
carried out by the authors in June and October, 1980 and in May,
1981. The search was made for astronomical alignments at three
sites. One such alignment has been found, and it is shown that
Müller's results, previously thought to be correct, differ considerably from the actual situation. In conclusion some general remarks
are made on the origin of the rings and their connections with the
migrations of the Goths.

THE SITES

Węsiory ($\phi = 54°13'.3$ N, $\lambda = 17°51'.0$ E)

This site was first mentioned in 1938 (Petsch 1938), but was not
excavated until the late 1950s (Kmieciński 1958-1968). It consists of 4 rings
(2 of which are incomplete) and 20 barrows situated on the shores of a picturesque
lake.

Fig. 1

<u>Odry</u> ($\phi = 53°54'.0$ N, $\lambda = 17°59'.7$ E)

The stone rings at Odry have been very well known for more than 100 years. Therefore we do not want to repeat everything that has been written about them. All that we can state here is that this cemetrey consists of 10 rings and 27 burial barrows, but there are many more graves with no recognizable marking on the surface.

Odry is best known, of course, because of P. Stephan's (1914)

Fig. 2

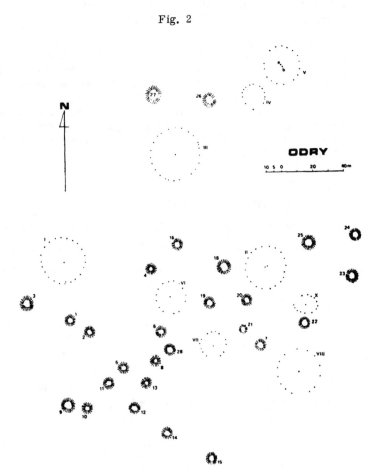

Fig. 3.   Odry: central stone in the ring III.

astronomical theory and more recent work by R. Müller (1934, 1970), where it
is pointed out that the structure of this site is oriented to the four cardinal points
as well as to the solstices. Müller rejected all previously supposed stellar alignments but one, i.e. the direction of Capella's setting, but he himself added one
new line, pointing to the setting/rising Moon in its extreme northern/southern
declination. These considerations resulted at last in a very logical structure.
All these conclusions, however, were drawn from Stephan's measurements,
refined later partly by Müller, and never questioned since then by any author for
about 50 years.

Grzybnica   ($\phi = 54°04'.2$ N, $\lambda = 16°27'.1$ E)

This site was found as late as 1975 (Wołągiewicz 1975, 1977), some
25 kilometres south of the city of Koszalin. Four rings and 11 barrows were
found there, but the site is not yet fully explored.

Fig. 4

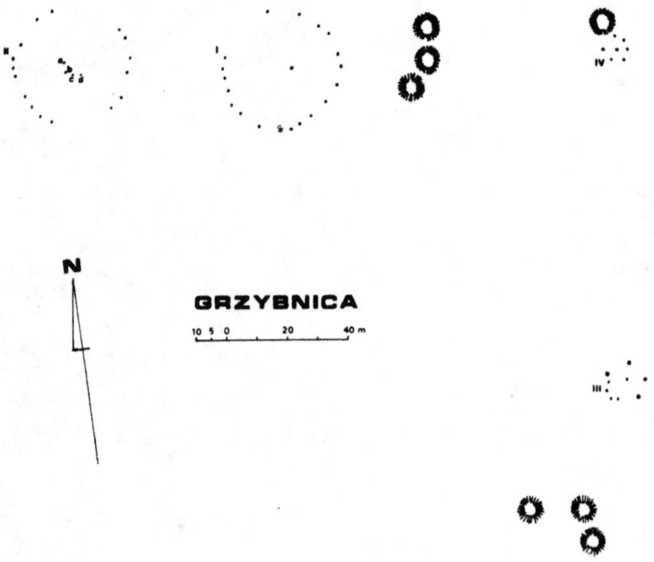

GRZYBNICA

OUR SURVEY

The method

Our intention was to make a set of homogeneous measurements at all three sites to look for possible similarities, and therefore we have treated all of them in the same way. The work has been carried out by the same team and with the help of the same theodolite. In order to obtain the true values of the measured azimuths we have found also the exact azimuth of one direction at each of two sites - Odry and Węsiory - by comparing it with the observed azimuth of Polaris. Unfortunately, we did not succeed in making similar observations at Grzybnica. That gave us a set of data which was quite independent of any previous results. From these azimuths the declinations were then calculated (using mean refraction), but for the mathematical horizon (h = 0) only, as the true horizon could not be observed from any of the sites because of lush vegetation.

The results

Fig. 5

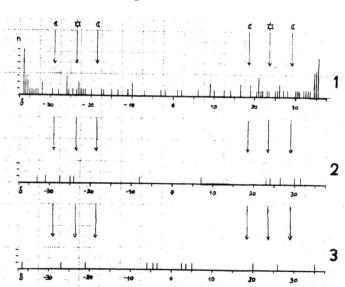

The final results of this procedure are shown on the diagram (fig. 5), where the declination is plotted with an accuracy of 0.5°, and the heights of strokes are proportional to the number of "hits" within this range.

The most surprising fact to emerge from our work was the systematic difference between our results and those of Müller (see Table 1). The mean deviation is as high as +2.43°, and therefore it is not to be neglected. We still do not know what is the reason for such a big difference, yet we have no doubts that our measurements were accurate.

This fact, in any case, changes the whole astronomical interpretation of the Odry site, for the accuracy of its main alignments falls down considerably, and one of them (the Moon in extreme declination) even disappears completely.

There is another misconception in Müller's theory, which is perhaps partly responsible for this situation. As we know, he was deeply convinced that the Odry rings were built about 1800 B.C.; therefore all his calculations were made for that epoch only. He had ignored some archaeological arguments, which were already known then and have since been proved to be entirely correct, that the rings were built approximately in A.D. 150. At that time in the territory of Pomerania lived people belonging to the so-called Wielbark culture. It

Table 1 Comparison of measured azimuths

From - to	Müller	Our data	Δ
III-Vb	49.55°	50.83°	+1.28°
I-V	48.1	51.07	2.97
I-III	46.40	49.65	3.25
III-II	138.95	141.68	2.73
VI-IX	136.4	137.4	1.0
III-I	228.1	229.65	1.55
VIII-III	326.5	330.0	3.5
IX-II	337.2	339.6	2.4
I-II	90.3	92.6	2.3
III-II	138.4	141.7	3.3

$$\overline{\Delta} = +2.43°$$

is now almost certain that this culture formed from a local substrate under the influence of Gothic groups coming there from Scandinavia.

It can be easily seen on the map (fig. 6) , where every dot represents a site with stone rings, that these structures were spread widely throughout southern Scandinavia, the homeland of the Goths. It should also, of course, result in a certain unity in the arrangement of the rings on the sites on Polish territory. Indeed, one can find the same rough solstitial alignment at Odry and Węsiory. The lack of that direction at Grzybnica may be due, perhaps, to its unfinished archaeological exploration.

At the present stage of research any other conclusion seems to be rather premature. As we have mentioned at the beginning, the problem needs further investigation and statistical analysis of the results. Consequently we plan to do this in the very near future, together with an attempt to implement the statistical test of Freeman and Elmore.

Acknowledgments

This paper has been prepared with substantial support from the Polish Archaeological and Numismatic Society and the Department of Historical Anthropology of the Institute of Archaeology, Warsaw University. Our thanks are also due to Drs J. Kmieciński, T. Grabarczyk and R. Wołągiewicz for their comments and for permission to carry out our research on the sites. In addition we wish to express our personal thanks to Mr Z. Skrok and Mr R. Kirilenko.

We would also like to acknowledge the most helpful assistance of Drs M. A. Hoskin and D. C. Heggie, as well as of the International Astronomical Union, whose efforts made it possible for this paper to be presented at the Archaeoastronomy Symposium in Oxford, 4-9 September 1981.

Fig. 6

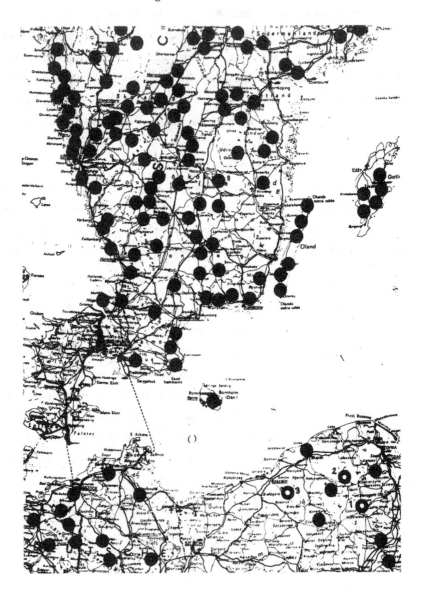

BIBLIOGRAPHY

This bibliography lists only a selection of titles connected with the archaeology and astronomy of the stone rings in Sweden, Poland, and Germany.

Almgren, O. and Nerman, B. (1914). Die ältere Eisenzeit Gotlands, Vol. 1-2, Stockholm.

Bohnsack, D. (1940). Ostgermanische Gräber mit Steinpfeilern und Steinkreisen in Ostdeutschland. Gothiskandza, 2.

Dworak, T. Z. and Kubiak, M. (1981). O mozliwości zaobserwowania heliakalnego zachodu Capelli w Kręgach Kamiennych w Odrach. Urania, no. 6.

Hachmann, R. (1970). Die Goten und die Skandinavien. Berlin.

Hirschfeld, G. v. (1877). Die Steindenkmäler der Vorzeit. Zeitschrift des historischens Vereins Marienwerder, 2.

Kmieciński, J. (1958). Niektóre zagadnienia wędrówki Gotów w świetle dotychczasowych badań oraz w świetle wykopalisk w Węsiorach w pow. kartuskim. Zeszyty Naukowe UŁ, Nauki Hum. -Społ., Seria I, 8.

Kmieciński, J. (1959). Niektóre społeczne aspekty epizodu gockiego w okresie środkoworzymskim na Pomorzu. Zesz. Nauk. UŁ, Nauki Hum. - Społ., Seria I, 12.

Kmieciński, J. (1962). Zagadnienie tzw. kultury gocko-gepidzkiej na Pomorzu Wschodnim w okresie wczesnorzymskim. Łódź.

Kmieciński, J. ed. (1966). Cmentarzysko kurhanowe ze starszego okresu rzymskiego w Węsiorach w pow. kartuskim. Prac i Mat. Muz. Arch. i Etn. w Łodzi, Seria Archeologiczna, 12.

Kmieciński, J. Ed. (1968). Odry - cmentarzysko kurhanowe z okresu rzymskiego w pow. chojnickim. Łódź.

Kostrzewski, J. (1928). Kurhany i kręgi kamienne w Odrach w powiecie chojnickim, na Pomorzu. Muzeum Wielkopolskie w Poznaniu, 3.

Kostrzewski, J. (1934). Zagadnienie tzw. kultury gockiej na ziemiach dawnej Polski. In Tydzień o Pomorzu. Poznań.

Kozłowski, L. (1921). Groby megalityczne na wschód od Odry. Prace i Materiały Antropologiczne, Archeologiczne i Etnograficzne, 2.

Kubiak, M. (1980). Astronomiczne próby ustalenia wieku Kręgow Kamiennych w Odrach. Urania, no. 11.

Kubiak, M. (1981). Astroarcheologia: dalsze uwagi i hipotezy na temat powstania Kręgów Kamiennych w Odrach. Urania, no. 7.

Labuda, G. (1968). O wędrówce Gotów i Gepidów ze Skandynawii nad Morze Czarne. In Liber J. Kostrzewski octogenario veneratoribus dicatus. Wrocław.

Lissauer, A. (1874). Die Cromlechs und Trilithen bei Odri. Schriften der Naturforschenden Gesellschaft zu Danzig, NF 3 (3).

Lissauer, A. (1887). Die prähistorischen Denkmäler der Provinz Westpreussen und der angrenzenden Gebiete. Leipzig.

Müller, R. (1931). Die astronomische Bedeutung des mecklenburgischen Steintantzes bei Bützow. Praehistorische Zeitschrift, 22.

Müller, R. (1933). Zur astronomischen Bedeutung des mecklenburgischen
    Steintantzes bei Bützow. Praehistorische Zeitschrift, 24.

Müller, R. (1934). Zur Frage der astronomischen Bedeutung der Steinsetzung
    Odry. Mannus, 26.

Müller, R. (1970). Der Himmel über dem Menschen der Steinzeit. Springer
    Verlag: Berlin.

Müller-Quales, G. (1940). Die Goten: Vorgeschichte der deutschen Stämme.
    Leipzig/Berlin.

Okulicz, J. (1970). Studia nad przemianami kulturowymi i osadniczymi na
    Pomorzu Wschodnim, Mazowszu i Podlasiu. Archeologia Polski,
    15.

Petsch, G. (1938). In Weichselland, 1.

Piekarczyk, S. (1963). O społeczeństwie i religii w Skandynawii VIII-XI w.
    Warszawa.

Schindler, R. (1940). Die Steinkreise von Odry. Der Deutsche Osten, 8.

Stephan, P. (1914). Vorgeschichtliche Sternkunde und Zeiteinteilung. Mannus,
    7.

Strömberg, M. (1961). Untersuchungen zur jüngeren Eisenzeit in Schonen.
    Acta Archaeologica Lundensia, vol. I and II. Lund.

Strömberg, M. (1963). Kultische Steinsetzung in Schonen. Meddelanden fran
    Lunds Univ. Hist. Mus. Lund.

Tylżanowska, A. and Potemski, H. (1963). Kręgi kamienne w Odrach.
    Bydgoszcz.

Weibull, C. (1958). Die Auswanderung der Goten aus Schweden. Göteborg.

Wildte, F. (1926). Tingsplatserna i Sverige under förhistorisk tid och
    medeltid. Fornvännen, 21.

Wołągiewicz, R. (1975). Grzybnica - cmentarzysko kultury wielbarskiej z
    kręgami kamiennymi. Koszalińskie Zeszyty Muzealne, 5.

Wołągiewicz, R. (1977). Kręgi kamienne w Grzybnicy. Koszalin.

Zajdler, L. (1980). Tucholski Stonehenge a Capella - pomorski Syriusz.
    Urania, no. 11.

ASTRONOMICAL ORIENTATION OF NEOLITHIC SITES IN
CENTRAL EUROPE

W. Schlosser
Ruhr-Universität, D-4630 Bochum, Germany
J. Čierny
Ruhr-Universität, D-4630 Bochum, Germany

Abstract. The orientations of more than 2200 finds
covering linear pottery, corded ware and bell beaker
are discussed. The azimuthal distribution is shown
to be characteristic for each culture. Furthermore,
the skeletons and/or graves were oriented towards
the cardinal points rather than actual azimuths of
the rising or setting sun. The accuracy of determing
true north is estimated to be of the order of 3°.
This can only be achieved by employing methods like
the Indian circle. Thus a catalogue of minimal knowledge of geometry and astronomy may be derived. The
anthropological data were used to construct a population curve.

Introduction

Since 1978, the Institutes of Astronomy and Prehistory
of the Ruhr-University Bochum collaborated in an interdisciplinary project to investigate neolithic orientations. Data on over
2200 finds have been collected and prepared for electronic data
processing. One of the aims was to establish a catalogue of
minimal knowledge in geometry and astronomy in neolithic times.
Initially, the well-documented Bohemian-Moravian bell beaker
and corded ware cultures were selected, because in two previous
papers U. Fischer (1953, 1956) showed their tendency for non-
uniform orientation. Since both cultures terminate the neolithic
epoch, in 1980 an early-neolithic culture - linear pottery - was
added to the program. Due to the relatively homogeneous cultural
pattern of linear pottery, data from a larger region were collected, ranging from Alsace to Slovakia.

In this paper, the main results of the project are
presented. For full details and references see Schlosser et al.
(1979 and 1981).

Azimuthal distribution of orientations

The orientations of the skeletons (skull to pelvis) and/or the major axes of the graves were derived from the published records. Depending on their accuracy, the subdivision of azimuths ranged from 1° (modern excavations) to 22.5^o (more poorly documented older excavations). The azimuthal distribution of orientations for the three cultures is given in figs. 1-3. To facilitate comparison, all three distributions are subdivided in the same manner (22.5^o = 1/16 of full circle).

Fig. 1. Orientation of linear pottery (about 5000 B.C.)

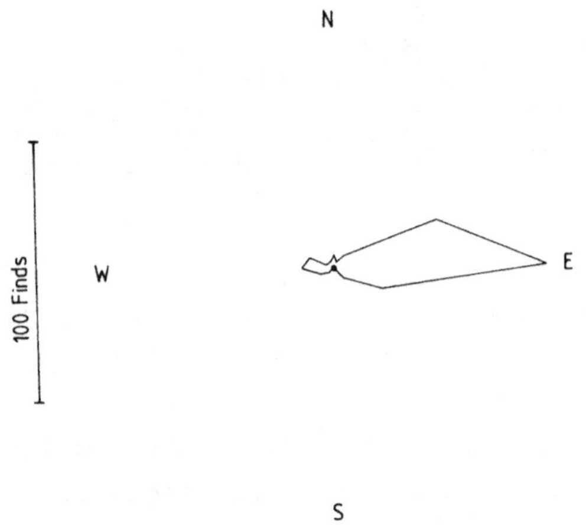

The differences between the cultures are clear-cut. Furthermore, the pronounced maxima towards the four cardinal points prove that neolithic man had already abstracted the ever changing azimuths of the rising or setting sun to "east" and "west". To clarify this, fig. 4 gives the distribution of all finds at Aiterhofen (Bavaria, linear pottery). The orientations cluster around 90° (east).

Fig. 5 shows a distribution of a _hypothetical_ population, which buries its members according to the actual azimuth of the rising sun at the day of death or burial. Because of the slower horizontal motion of the sun around the solstices compared to the equinoxes, a completely different angular distribution results with a minimum at 90° and two maxima at both ends. Fig. 5 has been drawn assuming a constant rate of mortality over the year. A more realistic approach would reduce the differences somewhat without changing the basic structure of fig. 5.

Accuracy of neolithic determination of azimuth and
its implications for Stone Age astronomy and geometry

Large sites with 20 graves or more mostly show a
"standard orientation", while the individual orientation of
graves at small sites is more irregular. Thus, large sites
preserve more truly the traces of neolithic astronomy and geometry. The large sites follow the general scheme valid for all sites (figs. 1-3) : there exist deviations from the cardinal points but at these points the density of finds is highest. For instance, three out of the six large sites with precisely determined orientations are directed towards east within 3° (Aiterhofen, Senghofen, Vikletice), while the remaining three differ widely (azimuths in brackets): Niedermerz (46°), Elsloo (115°) and Nitra (142°).

Fig. 2. Orientation of corded
ware (about 2200 B.C.)

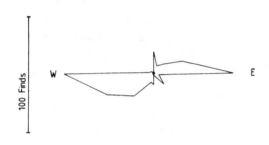

Fig. 3. Orientation of bell
beaker (about 2200 B.C.)

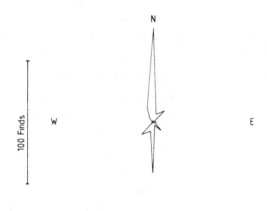

Elementary astronomy allows precise determination only of the north/ south direction using the upper or lower culmination of a celestial object. There is no simple method of determining east or west. So the first elements of the catalogue of minimal knowledge of astronomy and geometry are: (1) existence of an

invariable azimuth of culmination (fig. 3); (2) importance of
the right angle in geometry (figs. 1,2). The accuracy of de-

Fig. 4. Distribution of azimuths
at Aiterhofen as an example for
an east-oriented site.

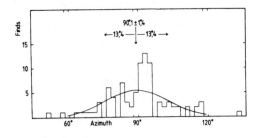

Fig. 5. Distribution of azimuths
for a hypothetical population using
the actual azimuths of the rising
sun.

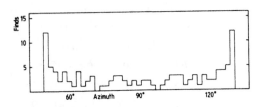

Fig. 6. Central part of fig. 4
for deriving the excess over the
Gaussian distribution as measure
for neolithic accuracy of azimuth
determination (hatched area).

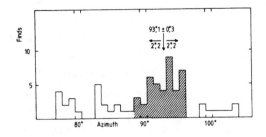

termination of azimuths
may be judged from the
systematic errors. In the
case of Aiterhofen (which
is representative for the
other east-oriented sites
not further discussed in
this paper) the accuracy
is estimated from the
slight shift of the ex-
cess over the Gaussian
distribution (fig. 6,
hatched area). It amounts
to about 3°.

A precision of 3°
cannot be achieved simply
by measuring the azimuth
of the shortest shadow of
a gnomon around noon. The
shadow does not change its
length significantly over
quite an azimuthal span
around this time. Instead,
the somewhat more sophis-
ticated principle of the
Indian circle had to be
employed. Basically, two
shadows of equal length
are measured before and
after noon. North then
lies on the bisector of
the connecting line
(fig. 7). So the catalogue
of minimal knowledge has
to be expanded: (3) know-

ledge of symmetry of the diurnal motion of the sun with respect to the azimuth of culmination; (4) property of the north-south direction to be the bisector of the azimuthal angles of both shadows and passing through the bisector of the connecting

Fig. 7. Geometry of the Indian circle.

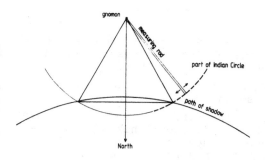

line. A static point of view would suggest a limitation to the geometry of the triangle. However, the triangle in fig. 7 cannot be drawn without the corresponding circle or, at least, parts of it. Either the Indian circle was drawn as a more or less complete circle before measurement or it was realized in two small segments by the circular motion of a measuring rod while determing the moments of equal lengths of shadows. The baseline of the triangle therefore is constructed as the cord of a circle. So one has to include in the catalogue of minimal knowledge: (5) knowledge of the circle and the chord.

Demographic results

For about 23 per cent of the finds anthropological information concerning sex and age was available. The age was subdivided on the well-known scale infantile-senile. Using the

Fig. 8. Population curves for both sexes of the three cultures.

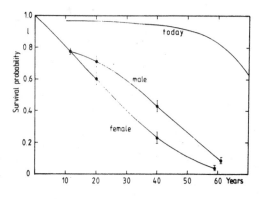

standard set of formulae describing the mortality rate of a stable population (e.g. Schlosser et al.,1979), population curves for both sexes were derived (fig. 8). No significant differences for the three cultures discussed in this paper were found. The systematic difference between male and female survival coefficients are attributed to the mor-

tality of women due to childbed fever. From fig. 8 a rate
of 5 % / birth can be deduced. Furthermore, the slope of the
female population curve is consistent with an average number
of about four children per woman.

References

Fischer, U. (1953). Die Orientierung der Toten in den neolithi-
    schen Kulturen des Saalegebietes. Jahresschrift für
    mitteldeutsche Vorgeschichte 37, p.49 ff.

Fischer, U. (1956). Die Gräber der Steinzeit im Saalegebiet.
    Vorgeschichtliche Forschungen 15, Berlin.

Schlosser,W., Mildenberger, G., Reinhardt, M., Čierny, J. (1979).
    Astronomische Ausrichtungen im Neolithikum. Ein Ver-
    gleich der böhmisch-mährischen Schnurkeramik und
    Glockenbecherkultur. Ruhr-Universität Bochum.

Schlosser, W.,Čierny, J., Mildenberger, G. (1981). Astronomische
    Ausrichtungen im Neolithikum. Ein Vergleich mittel-
    europäischer Linienbandkeramik. Ruhr-Universität
    Bochum.

STONE CIRCLE GEOMETRIES : AN INFORMATION THEORY APPROACH

J.D. Patrick
Deakin University, Victoria, 3217, Australia
C.S. Wallace
Monash University, Clayton, Victoria, 3168, Australia

Abstract. This article discusses the techniques of A. Thom
in deriving geometric designs to fit stone circles and from
this background argues for an alternative definition of an
hypothesis in scientific research. The definition that is
advocated herein is a union of Solomonoff's application of
Information Theory to inductive inference, Wallace's
Information measures and Halstead's software science
measures. This approach is applied to the comparison of
Thom's hypothesis against the authors' hypothesis that stone
circles are meant to be roughly circular and locally smooth
to the eye. The authors' hypothesis is modelled by a fourier
series wrapped around a circle. The results from 65 Irish
sites show that the authors' hypothesis is favoured at odds
of better than 780:1 compared to Thom's hypothesis.

INTRODUCTION

The stone circles of Britain have undergone detailed study
and statistical analysis over the past decade. The progenitor of this
work, Professor A. Thom, claims that these monuments are set out to
accurate geometric designs with the use of a standard unit of length,
the 'megalithic yard' (MY) equal to .829m or 2.72 ft (Thom 1967). This
claim has been investigated in two recent statistical analyses which
both concluded that only the stone circles from Scotland lent some
support to Thom's theories (Kendall 1974; Freeman 1976). Both these
studies as well as Thom's used about 200 circles whose diameters were
estimated by Thom. The statistical examination of a population of
diameters for evidence of quantisation is very difficult, as the
diameters are not basic data, but are inferred, with uncertain error and
bias, from the surveyed positions of individual stones. It is therefore
not surprising the previous analyses have produced very tentative
conclusions.

1   WHY MEGALITHIC SCIENCE NEEDS A NEW APPROACH

In Thom's studies there are two underlying assumptions. The
first assumption is that the geometric design that he has devised for
each site is "correct" in some sense. Thom indicates in his writings

that he tests a number of designs on a site plan from which a solution is accepted as "the best". Of course, in this context "the best" is poorly defined. Indeed, Angell (1976, 1977) has been critical of this deficient approach as there are no firm criteria by which to select the most appropriate solution. Kendall (1974) felt that analysing the circle diameters is a safe task stating that at least "a circle is a circle". However, he felt that the dimensions of non-circle sites were open to debate of design choice or unconscious bias. Freeman (1976), who also analysed Thom's data, voices stronger misgivings in saying "I disagree, of course, with Thom's claim that we know the design that was originally used".

The second assumption is the selection of a single site dimension to use in an analysis for an underlying quantum. There has been no discomfort expressed publicly by any commentators in using the diameters of circles as an appropriate dimension for analysis, apart from Angell (1976). However Kendall, Freeman and Angell have declared misgivings at using Thom's dimensions from his non-circle designs.

The analyses of both Kendall and Freeman were constrained to work with the data supplied by Thom, that is the diameters of some 200 sites. Thom himself no doubt feels that as he has justified the primary assumption, that is, the choice of geometric design applicable to a site, then no more statistical validation of shape selection is either necessary or appropriate. Thus he embraces the statistical results yielded by the Broadbent tests without equivocation (Thom, 1967), which of course support the existence of the megalithic yard.

The approach in this paper differs from previous analyses, as a fully specified Thomsian hypothesis with specified quantum is compared against our own version of a smooth non-quantal hypothesis. Kendall's analysis took the quantum as unspecified, and Freeman's analysis was designed to estimate the value of the quantum on the assumption that a quantum was used.

It is necessary to construct formally the details of Thom's hypothesis and so eradicate any unjustifiable assumptions, or at least make them the same as the assumptions underlying the hypothesis against which Thom's will be later compared. In attempting to compare Thom's hypothesis (H_T) against the authors' alternative (H_p) some details of H_T had to be inferred from Thom's published analyses, as Thom himself has not to our knowledge yet published a statement of his hypothesis sufficiently precise for testing. We believe our formalisation of H_T represents a fair summary of the state of Thom's work on stone circle geometry up to the present, but concede that it includes some assumptions not inferred from Thom's analyses. These assumptions concern matters not inconsistent with his work. An example is our assumption that the measured radial distances of stones from the centre of a circular site have a Normal distribution around the nominal radius.

The first and perhaps most obvious element of Thom's hypothesis is the definition of all the geometric designs Thom considers were used by the stone circle builders. The second element is the number of different shapes available in the complete retinue of shapes possible. Thirdly, Thom's solutions provide us with the frequency of use of the different designs within any region. These last two elements are very much a function of the stone circles surveyed to date, that is the evidence that has been collected so far. The importance of this fact will be demonstrated further on.

The analysis by Thom of each surveyed site has led him to formulate specific properties of stone circle designs. Firstly, that there was in use a standard unit of length equal to .829m. Secondly, that multiples and sub-multiples of this unit were used in a variety of ways. Thirdly, that there were special dimensional relationships in the design of ellipses, flattened circles and eggs, for example, the use of the pythagorean triple relationship for fundamental design elements.

In an analysis of Thom's theories one would like to eliminate all the underlying assumptions or explicitly incorporate them into the structure of an examinable hypothesis. However, it is also necessary to overcome the objections of Angell (1976) that site dimensions cannot be assessed accurately. Thom's plea that his work in toto must be evaluated needs to be satisfied (see discussion in Freeman 1976), though he has never done this himself. Satisfying these pleas means incorporating assessments of not only his quantum, but the perimeter conditions, the different geometric shapes, the use of pythagorean triads and the inherent inaccuracies in the data.

A statistical technique has been devised to fulfil all these requirements. However, there are a number of important facets of this technique that need to be emphasised before its implementation is described. Firstly, this technique can only compare hypotheses. Any

number of hypotheses can be ordered in terms of their efficiency at
describing a given set of data or evidence, but it does not and cannot
prove an hypothesis. Secondly, the comparison is solely based on the
evidence available, which in this case is principally the positions of
stones at each site. However, further evidence such as the inherent
geographical and/or archaeological distribution of sites could be
incorporated as evidence if one so desired. Thus the results from this
technique are wholly dependent on the evidence available and should new
evidence come to light the preference of one hypothesis over another may
be reversed.

2   THE COMPETING HYPOTHESES
         The archaeological background suggests that the
megalithic traditions were started by essentially isolated egalitarian
farming communities (Burgess 1974, Burl 1976). Over long periods they
developed practical engineering skills but each region retained its
individual architectural styles often regulated by local building
materials. It is plain that the visual element in all the large
monuments is the most important architectural feature, exemplified by
wide facades to tomb entrances, sometimes lined in quartz, e.g.
Newgrange, the high mounds that enclose superbly corbel vaulted tombs
and the long avenues of stones and banks that form entrances to
Stonehenge and Avebury. In this context we don't believe there was a
specific plan view held by the architects but only a ground view and so
generally no specific and detailed ground plan of a site was ever
formulated.

         Our model is based on our belief that the only
specific geometric shape intended by the builders was the circle, and
that other clearly non-circular shapes arose as informal modifications
to a circle intended to accommodate such features as a flattened facade
(in the tradition of passage graves), or elongation along an axis of
bilateral symmetry as seen in the Cork-Kerry sites (Barber 1972) of
Ireland. Our families of curves are not intended to denote any
typological sequence but rather the range of shapes that arise when the
plan geometry of sites is a matter of expediency and local smoothness.

         The set of functions available under H_T is the
union of several parameterised families, e.g.

$t_0(\theta,R)$:   circles of radius R,

$t_2(\theta,a,b,\beta)$:   ellipses with semi-major axis a,

semi-minor axis b and

orientation β,

$t_3(\theta,R,\beta)$:   flattened circle of radius R

and orientation β,

etc.,

where the parameters conform to integer and geometric constraints which
Thom has described, and where the expected relative frequencies of the
different families can be inferred from his analyses of other sites,
circles being the most abundant shape.

Similarly, our H_p is of the general form

$$f(\theta) = R + a_2 \sin 2\theta + b_2 \cos 2\theta + a_3 \sin 3\theta + b_3 \cos 3\theta$$

but for convenience we divide the shapes into the following families;

$f_0(\theta,R)$: circles of radius R,

$f_2(\theta,R,a_2,b_2)$: shapes with second-order Fourier

terms,

$f_{3S}(\theta,R,r_2,r_3,\beta)$: shapes with second- and third-order

terms with bilateral symmetry of

orientation β,

$f_3(\theta,R,a_2,b_2,a_3,b_3)$: shapes with second and third order

terms,

these families being listed in decreasing order of expected abundance.
f_{3S} is a specialisation of f_3 obtained by putting,

$$a_2 = r_2 \sin 2\beta, \quad b_2 = r_2 \cos 2\beta, \quad a_3 = r_3 \sin 3\beta, \quad b_3 = r_3 \cos 3\beta.$$

Examples of these shapes are shown in Figures 1 and 2, where
A is used in place of θ to denote the angle in the polar equation.

3   MEGALITHIC STUDIES TO DATE - ENGINEERING DESIGN

The Thoms and their supporters have completed a great many
surveys which constitute a very large body of data. However, at times it
is unclear what the exact problem is to be solved, what criteria for
data collection have been applied in terms of the accuracy of the data,

exactly what data is collected and the manner in which the raw field
data should be transformed to make it meaningful for the problem solving
processes.

   Certainly a great deal of ingenuity and imagination has been
well used in the development of suitable solutions for the design of
megalithic sites. The designs created by the Thoms and Cowan (1970) and
the like lack nothing in perceptiveness. Some iterative testing of
solutions has apparently been performed though there is no definitive
picture of any systematic or exhaustive comparisons. Plainly, the large
range of geometric shapes advocated by Thom indicates many hours of
experimentation. However, it is on this point that Thoms' engineering
methodology and general scientific methodology are manifestly opposite.

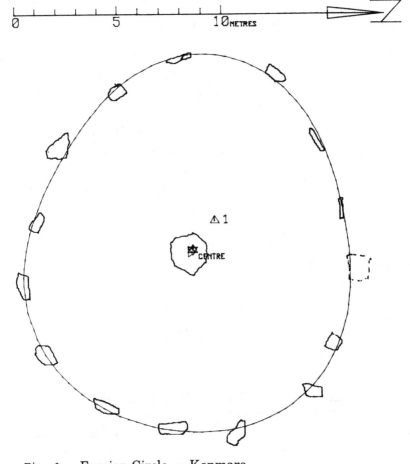

Fig. 1   Fourier Circle — Kenmare

$$F_{3s} = 8.22 + .48 \cos2(A+.2) - .41 \cos3(A+.2)$$

The interactive testing of solutions to the problem, i.e. the design
of megalithic sites, is regulated by criteria applied subjectively, as is
often the case in engineering design. The design solutions for many
sites published by the Thoms have been selected on intuitive grounds, on
the appeal of the implausibility of alternative solutions or on
goodness-of-fit calculations or the circumstantial evidence of
comparisons with other sites for which a design solution has previously
been published. None of these selection criteria are acceptable under
scientific methodology.

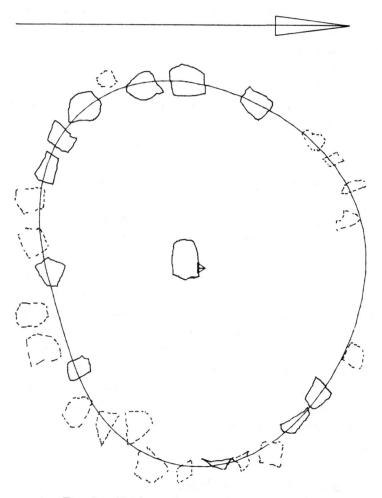

Fig. 2   Fourier Circle — Carrowmore 04

$$F_{3s} = 5.88 - .47 \cos2(A-1.9) + .32 \cos3(A-19)$$

The application of the scientific method requires the systematic tracing of five steps. The first step is to define the reason for the research and what it is supposed to achieve. This step is quite similar in engineering design. Secondly the modus operandi must be defined in terms of the appropriate research strategies and techniques. In particular it is necessary to decide if one is operating on a deductive strategy, that is to collect data for comparison with an existing hypothesis or an inductive strategy, that is to formulate a new hypothesis.

The third step is to direct the inquiry so that relevant and sufficient evidence is collected for the analysis. Fourthly, the analysis is performed and the outcome is clearly stated either in the form of a new hypothesis (induction) or the extent of the evidentiary support (deduction). The final step and the most important stage for anyone external to the project is the documentation of what was done, what was found and the significance of the findings (Buckley et al.1976).

The current state of the study of megalithic geometry and astronomy is at the inductive inference stage. There has been collected a large number of particular instances and facts and one must move into a tentative generalisation which seems to comprise them all. Many people will complain that this has been done but in fact it has only been done in the engineering sense in that solutions have been supplied within the framework of somewhat arbitrary criteria. The inductive inference has not been completed in the scientific sense as proper testable hypotheses have not been explicitly formulated. As a consequence of this failing the probabilistic analyses of the past have fallen short of shedding much light on the design features of stone circles. For those who consider the inductive stage has been comprehensively fulfilled let them go to any site not previously analysed and apply their deductive reasoning to their expectations of the Thomsian design of that site. They will find very quickly that they have no explicit criteria whatsoever as to the most appropriate astronomical or geometric design for the site.

4   AN HYPOTHESIS - THE CURRENT DEFINITION

For an hypothesis to be good it must fulfil the following criteria (Emory, 1980):

    - adequacy,i.e. it must clearly state the condition,size or

distribution of some variable or variables in terms of values
meaningful to the research task.

- testable, i.e. an hypothesis is untestable if it requires
techniques which are unavailable with the present state of
the art.

- better than its rivals, i.e. it covers a greater range of
facts but is simple in requiring fewer conditions or
assumptions.

This last point emphasises the need for an investigator when
formulating an hypothesis to find a balance between complexity and
simplicity. The Thoms' work falls short on the first of these
criteria and its conformity to the third criterion is a major area of
debate.

Once a good hypothesis has been formulated and appropriate
data collected it is a standard statistical approach to formulate an
opposite or null hypothesis. The texts on research methodology
emphasise that hypotheses are not proven, but they do say that the
statistical tests enable one to accept or reject the original
hypothesis. Thus one is provided with a quasi-proof in a manner not
unlike the acceptance of an engineering design after the testing and
analytical stages.

In this framework of the acceptance or rejection of an
hypothesis there is one situation that can cause considerable
difficulty. That is the situation where there is no clearcut support
for either the hypothesis or its null version. At this point, one can
only resort to the dictum of Occam's Razor that "multiplicity ought not
to be posited without necessity" (Enc Brittanica).

5   HYPOTHESIS - A NEW DEFINITION

The real requirement of the scientific method is a measure,
on a continuous scale, of the extent to which the evidence supports the
hypothesis. A continuous measure of evidentiary support would permit
the use of a multiplicity of hypotheses and an effective ranking of
each. By inference the highest ranking hypothesis would be the most
probable explanation of the data and the ranking of hypotheses could
quite easily change should new evidence come to hand. The current
approach of accepting or rejecting an hypothesis probably extends from
our desire to know the 'real' answer, which is not really possible.

This attitude parallels the engineering approach of settling on a specific design or solution in preference to any other.

To create this continuous measure it is necessary to develop axiomatically some new perspectives on the nature of an hypothesis.

AXIOM 1
An hypothesis is a proposition that purports to describe a pattern or order in a set of observable data.

The principle of coding theory is that any message can be coded into another shorter message if there is any pattern in the symbols used in the original message.

Thus a deduction from this axiom is that an hypothesis is an explanation of observable data that purports to describe the data in a message briefer than a full itemisation of the data set itself. Such an explanation is an encoding algorithm and therefore can be assessed of its merit at describing the pattern in the data by the length of the message it generates to describe the data.

AXIOM 2
The aim of the scientific method is to discover the longest pattern sequence in a data set.

This activity is the process of developing more sophisticated hypotheses. If the same pattern or order is found in many data sets and the predictability of that pattern becomes useful for a wide range of deductive analyses then that pattern will become a Law of Nature.

By deduction from the two axioms it can be concluded that an hypothesis that gives a more comprehensive description of the pattern in a data set than its competitors will encode the data into a shorter message than the other hypotheses.

AXIOM 3
The complexity of the statement of an hypothesis is a function of the entities in the hypothesis and their relationships.

An hypothesis statement consists of entities and relationships. However the entities in an algorithm are the operands and the relationships are the operators. As an hypothesis is an algorithm the complexity of an hypothesis' statement is measured as a function of the operators and operands of the algorithm that encodes the data.

The complexity of an hypothesis can therefore be measured as the length of a program coded in some computer or pseudo-computer language. This deduction and the necessity to quantify an hypothesis for the application of Occam's Razor suggests an important corollary.

COROLLARY 1

To compare the complexity of two hypotheses they must be written in a language that does not favour the encoding of either hypothesis by its intrinsic data structures or operators.

The complexity of an hypothesis may be increased to explain data that deviates uncomfortably from the original hypothesis. This expansion creates the appearance of increased evidentiary support for the hypothesis. This is the situation with Thom's work. Alternatively, an hypothesis may be simple and so some of the highly deviant evidence may appear to be inadequately explained.

From the axioms and corollary set down previously and in satisfaction of Occam's Razor it is possible to deduce a suitable measure of evidentiary support.

COROLLARY 2

A continuous measure for the ranking of hypotheses and their evidentiary support is the length of the message that describes an hypothesis statement and the evidence (data) optimally encoded according to that hypothesis.

6   MESSAGE STRUCTURE

In Section 5 it has been established that an hypothesis is a description of a pattern in a data set and therefore can be used to encode the data into a message. It is important that the coding should be optimal in that it should produce the shortest possible message for each hypothesis that is to be compared.

The basis of optimum encoding is Shannon's Information Theory as applied in Huffman coding. Given the probability of each event in a string then Huffman coding enables one to produce the minimum message length for that string. Huffman coding has traditionally been used in computers for binary encoding and thus uses base 2 logarithms and so the length or cost of any message is in units of bits. In the context of this paper an hypothesis is also a predictor of the probability of each

event in the message. As the method compares the messages generated by
various hypotheses the logarithm base is immaterial and it is more
convenient to use natural logarithms. The units are therefore called
natural bits or nits. The message is called the "Information Measure"
and denoted I_p for hypothesis H_p and I_T for hypothesis H_T
(Wallace & Boulton 1968).

The result of the coding activity is the minimum message for
each hypothesis and the data. In applying the inverse of Huffman coding
the relative probabilities of hypotheses are readily determined from the
minimum message lengths.

The total message in any hypothesis consists of two principle
sections. They are the fixed overhead of describing the hypothesis and
secondly the variable cost of the encoded observational data. Each of
these sections may be broken into many components. In the stone circle
problem the first section consists of two components:

I1   – is the description of the hypothesis, i.e. a statement
of the Fourier family for H_p and for H_T a
description of the various geometric designs with
their rules of constraint;

I2   – the relative abundances of the various shapes. This
component might well be considered as part of the
hypothesis but it has been separated to permit a
certain flexibility. The relative abundances are
considered as parameters of a group of sites and
therefore are estimated from the data. Thus it is
assumed in both hypotheses that different collections
of sites might yield different relative abundances
of their various shapes.

The second section of any message is the description of the
data, which in this case is the positions of stones encoded according to
some geometric design. In this instance, each site message can be
divided into three components.

I3   – is the particular geometric shape by which the site
is described;

I4   – the description or value of each parameter of the
assigned geometric shape;

I5   – the description of each stone with respect to the
parameters of the assigned geometric shape.

7   DETERMINATION OF THE COMPLEXITY OF AN HYPOTHESIS-I1

The complexity of an hypothesis as defined in Axiom 3 and
Corollary 1 is the I1 component of the Information Measure. This
complexity is determined by consideration of a Turing machine.

The Turing machine fulfils the important criterion of not
having any intrinsic features that favour any of the competing
hypotheses in megalithic studies. Unfortunately to write the necessary
programs for Turing machines is a daunting task. However, the study of
the structure of programs written in conventional computer languages
offers assistance in this matter.

Earlier in this paper it was deduced from an axiom that the
complexity of an algorithm is a function of its operators and operands.
Halstead (1977) has demonstrated empirically in well-written
programs that there is an internal consistency between the number of
operators and their frequency.

If η_1 and η_2 are the unique operator and operand
counts respectively in a program and N is the total usage of all
operators and operands then

$$N = \eta_1 \log_2 \eta_1 + \eta_2 \log_2 \eta_2 \qquad (1)$$

Also Halstead derived a function for the volume of a program, i.e. the
total length of the program in bits,

$$V = N \log_2 \eta \qquad (2)$$

where $\eta = \eta_1 + \eta_2$.

The term $\log_2 \eta$ is the number of bits needed for the
unique definition of each operator and operand. The volume V of any
algorithm will be a function of the language in which it is written and
therefore it is necessary to design a special language for the current
needs. If hypotheses are to be compared on an equal footing then the
language they are written in should have no intrinsic features that
offer a description advantage to any one hypothesis. As well,
Solomonoff (1964) says of von Heerden's work that the arbitrary choice
of language to describe operators is equivalent to an arbitrary choice
of universal Turing machine.

Decoding algorithms for Thom's (H_T) and Patrick's (H_P) hypothesis on the design of stone circles were written in an ALGOL-like language. The language contains the usual arithmetic and logical operators, arrays, the facility for function and procedure definitions and calls, a block structure, FOR loop and CASE statement. The actual decode procedures for uniform and normal distributions are intrinsic to the language. Thus the difference between the two programs is essentially only the difference between the two hypotheses.

Table 1. Halstead's program parameters for H_P and H_T.

	H_P	H_T
n_1	31	40
n_2	71	99
N_1	289	424
N_2	195	326
V(bits)	3229	5339
λ	59.0	63.9

The H_P program decodes for all four families of shapes, i.e. f_0, f_2, f_3, f_{3S} where the H_T program decodes for t_0 (circle), t_2 (ellipse) t_{3A} and t_{3B} (flattened circles). The counts of operators and operands were made for both programs and the results are presented in Table 1. It can be seen that the H_T hypothesis is 2110 bits more than H_P, demonstrating clearly that H_T is by far the more complex hypothesis. The term λ is the language level and is a measure of the sophistication of the language. Typically ALGOL programs yield values between 2 and 4.

The very high values of λ for H_P and H_T indicate that the aim of writing in a language highly suited to the task has been achieved. The fact that λ for H_T is 4.9 higher than H_P shows that the program language is more suited to H_T. If it had been possible to write the programs so that the language levels were identical then the differential in program volume, and therefore hypothesis complexity, between H_P and H_T would be increased even further. It appears that the provision of arrays as a data structure in the programs has assisted the definition of H_T.

8   THE INFORMATION COST OF GEOMETRIC SHAPES - I2 & I3

The abundance of each geometric shape is really a component
of the hypothesis overhead i.e. I1. However to permit the flexibility
of determining the optimum relative abundances for H_p and H_T it is
separated computationally. This relative abundance, i.e. I2, is a
multi-state attribute of an entire data set that specifies the frequency
of occurrence of each possible shape. For both H_p and H_T there are
four permissible shapes. The information cost for a multi-state
attribute has been derived in a taxonomic context, by Wallace & Boulton
(1968), as the relative abundance of classes.

The message describing each stone circle must be prefixed by
the code specifying the particular geometric shape by which the site is
described. This piece of code, i.e. I3, is the shape label cost or in
the taxonomic sense the cost of specifying the class membership of the
site. It has been shown (Wallace & Boulton 1968) that the I2 and I3
components of a message can be combined for computational convenience.
The sum of I2 and I3 is derived to be

$$\frac{(T - 1)}{2} \; ln \; (S/12 + 1) - ln \; (T - 1)! - \sum_{t=1}^{T} (n[t] + 1/2) \; ln \; p[t]$$

where T is the number of states (permissible shapes),

S is the sample size (no. of sites),

n[t] is the number of data items (sites) assigned to state t,

p[t] is estimated by n[t]/S, i.e. the frequency of use of state T.

9   THE OPTIMUM INFORMATION COST OF A SITE - I4 and I5

I5 describes the positions of the stones using a code which
would be optimum were the circle indeed set out according to the shape
whose family is given in I3 and whose parameters are given in I4. The
measured positions of any stone can be specified in polar coordinates
as (r_i, θ_i). As neither H_T nor H_p makes any statement about
the distribution of θ's, the part of I5 giving the θ's is assumed
identical under both hypotheses, and its length is not computed. I5
therefore need only specify for each stone the difference between the
measured value r_i and the expected radius $g(\theta_i)$. If these
differences are distributed as $N(0, \sigma^2)$ the I5 message length is
approximately proportional to $\log \sigma$.

Thus, if a particular geometric shape family has many parameters (e.g. f_3 with 5 parameters), the I4 component for a site assigned to that family will be long, since I4 must specify the value of each parameter. However, one would hope that, by optimum choice of the many parameters, the radial discrepancies could be made small, so reducing the value of σ for the site, and hence reducing I5. On the other hand, if a site is assigned to a simple shape family, I4 will be short, but a poor fit may make I5 longer. Thus there is a trade off between the two components that produce a minimum IM. The full derivations are presented in Patrick (1978) for the case of each geometric shape. However, a simplified model is discussed below.

Consider the program which results if we restrict ourselves to simple circles, the centre of the circle is assumed to be already known and Thom's hypothesis is restricted to integer values for the radius. In this case, the data comprises, for N stones, an ordered set of N values (r_i). An "explanation" of the set of radius data will, under either H_T or H_P, take the form of a message with the following structure:

$$(\sigma) \quad (R) \qquad\qquad (r_1)(r_2)(r_3)...(r_i)...(r_N)$$

Preamble (I4)   Body (I5)

The preamble states the standard deviation and mean of the Normal distribution assumed for the radial positions of the stones and forms a simple "hypothesis" about the distribution of the stones. The body gives in turn the value of each radial distance, using a code which would be optimal if indeed the N distances were N independent random values drawn from a $N(R,\sigma^2)$ distribution.

The optimum code for independent random values drawn from a distribution employs a long encoding for improbable values and a short encoding for probable values. The expected length of a message using the optimum encoding for a value of probability p is $(-\log_2 p)$ bits or $(-\ln p)$ nits.

Generally the probability of a stone's position r_i can be expressed as

$$\int_{r_i - \delta/2}^{r_i + \delta/2} \frac{1}{\sqrt{2\pi}\sigma} e^{-(r - R)^2/2\sigma^2} dr$$

where δ is the accuracy of the survey measurements, estimated to be .01 m. Thus applying Shannon's Theory for optimum encoding and summing over N stones the message length for describing all the stones of a site is

$$I5 = -N \ln (\delta/\sqrt{2\pi}\sigma) + \sum_{i=1}^{N} (r_i - R)^2/2\sigma^2 \qquad (1)$$

Now let us say that the radius R can have any value up to a maximum L_R of, say, 35 my for both hypotheses H_p and H_T. Under H_T, the description of R is an integer in the range 1 to 35. Assuming for simplicity that all radii in this range are equally probable a priori, the length of this description is simply (ln35), whatever the value of R.

Under H_p, we can in I4 assign any value to R, but only to a limited precision, since the length of the message must be finite. If we decide to quote R to a precision or least count of U_R, then the I4 description of R has length $\ln(L_R/U_R)$. For instance, if we choose to quote R to the nearest 0.01 my, the description has length $\ln(35/0.01)$ or $\ln(3500)$, there being 3500 different possible values.

Similarly, under both H_T and H_p, we may assume σ to take values up to some limit L_σ, and if we decide to quote σ to precision U_σ the message length required is $\ln(L_\sigma/U_\sigma)$. Thus

$$I4 = \ln(L_R L_\sigma) - \ln(U_R U_\sigma) \qquad (2)$$

Now L_R and L_σ are common and identical for both hypotheses and so will make no contribution to discriminating between the hypotheses and shall not be considered further. Note that this is not generally applicable to all parameters for all geometric shapes.

It is necessary to derive expressions for R and σ that minimise the IM given the restraint that the quoted parameters in the message must be one of the set defined by U_R and U_σ. If U_R

is very small then the number of values available to be quoted is very
large and so I4 is long and costly. However, a dense set of R values
would make it cheap to describe each of the individual stone positions.
If U_R is very large then I4 would be cheaply described but the
stones may be expensive to encode. Thus as with the definition of the
optimum number of parameters there is also an optimum uncertainty for
each parameter that minimises the IM.

Whatever the model for the radial distribution of stones, the
optimum encoding of their positions on the assumption of a particular
intended contour (i.e. perimeter of the geometric shape) has a length
which is minus the logarithm of the probability of finding the positions
observed, given the assumed intended contour. Thus the estimates for
R and σ which minimise I5 solely are the maximum likelihood
estimates. However, because R and σ are stated only to limited
precision in I4 (a precision of 1 my in the simplified case of R under
H_T), the length of I5 using these values will on average exceed the
value obtained with maximum likelihood estimates.

Suppose, for a message using H_p, we quote R to precision
U_R and σ to precision U_σ. Let (R_0, σ_0) be the values
that minimise the I5 component of the message, and let (R_M, σ_M)
be the values actually quoted and therefore minimise I4+I5. Define R_M
$= R_0 + \Delta_R$, $\sigma_M = \sigma_0 + \Delta_\sigma$, where $|\Delta_R| \le U_R/2$
and $|\Delta_\sigma| \le U_\sigma/2$. We assume

$$E(\Delta_R^2) = U_R^2/12, \quad E(\Delta_\sigma^2) = U_\sigma^2/12 \tag{3}$$

Omitting constant terms from (1) and (2)

$$
\begin{aligned}
I4 + I5 &= N \ln\sigma + \Sigma_i (r_i - R_0)^2/2\sigma^2 - \ln (U_R U_\sigma) \\
&= N \ln\sigma + (v^2 + N(R_0 - m)^2)/2\sigma^2 - \ln (U_R U_\sigma)
\end{aligned}
\tag{4}
$$

where N is the number of stones and

$$m = \frac{1}{N} \sum_{i=1}^{N} r_i \ , \quad v^2 = \sum_{i=1}^{N} (r_i - m)^2$$

For any σ, (4) is minimised with respect to R_0 when $R_0 = m$, so we set

$$R_0 = m, \quad R_M = m + \Delta_R \tag{5}$$

giving

$$I4 + I5 = N \ln\sigma + (v^2 + N\Delta_R^2)/2\sigma^2 - \ln(U_R U_\sigma)$$

The precise value of Δ_R will depend on irrelevant details such as the origin chosen for the R_M scale. We therefore substitute the expectation of Δ_R^2:

$$E(I4 + I5) = N \ln\sigma + (v^2 + \frac{N}{12} U_R^2)/2\sigma^2 - \ln(U_R U_\sigma)$$

This is minimised with respect to U_R when

$$U_R^2 = 12\sigma^2/N \tag{6}$$

giving a minimum value

$$E(I4 + I5) = (N-1) \ln \sigma + v^2/2\sigma^2 + 1/2 - \ln U_\sigma - \ln\sqrt{12/N}$$

which is minimised with respect to σ when $\sigma^2 = v^2/(N-1)$.

We therefore set

$$\sigma_0^2 = v^2/(N-1), \quad \sigma_M = \sqrt{v^2/(N-1)} + \Delta_\sigma \tag{7}$$

Substituting its expectation for Δ_σ^2, and minimising with respect to U_σ gives (to second order in U_σ/σ)

$$U_\sigma^2 = 6\sigma^2/(N-1)$$

and

$$E(I4+I5) = -\ln(U_R U_\sigma) + N \ln \sigma_0 + \sum_{i=1}^{N} (r_i - R_0)^2/2\sigma_0^2 + 1$$

The interpretation of the 1/2 nit correction term for each parameter is that the message length is computed using the parameter estimates that minimise only I5. This value must always be at least equal to but usually smaller than the true length of I5 which uses the parameter value actually quoted in a message of minimum length. On average the necessary correction to I5 is expected to be 1/2 nit for each parameter.

Under H_T, the total computable message length is denoted as I_T and there is of course no question of choosing a precision for quoting R, as R can only be an integer in MYs. The length of the preamble for stating R is the logarithm of the number of integer MY values in the assumed range of 35 my. The best integer MY radius is simply the one closest to the mean radial distance, and the best value for σ is found by equating $N\sigma^2$ to the variance of the radial distances about the chosen integer radius. The optimum precision for quoting σ is again close to the expected estimation error.

The computation of the IMs is in practice complicated by several other factors, including the need to estimate the centre of the circle, the possibility of half-integer radii under H_T, allowance for the probable displacement of stones, particularly fallen ones, from their original positions, and of course the existence, under both H_P and H_T, of several different parameterised families of shapes available for the explanation of stone circles. These factors are all incorporated in our analysis and are discussed in more detail in Patrick (1978).

10   DATA GROUPS

The data used to compare the two hypotheses are drawn from four distinct regional groups of sites. The first group is the Cork-Kerry recumbent stone circles of southwest Ireland. Two previous studies of this group have given detailed archaeological assessments and descriptions (O'Nuallain, 1975) and an investigation of the sites' astronomical orientations (Barber, 1973). There are 79 extant sites in this region and 38 have been surveyed by the author (J.P.), but only 35 are suitable for numerical calculations, as the other three are too badly destroyed. The second group of 14 sites are from the Carrowmore passage grave cemetery in western Ireland about 10 km southeast of Sligo. Originally there were over 100 sites spread over a few square

kilometres but many of them are now destroyed. These sites do not fit
comfortably in the passage grave tradition as no actual passages have
been found. However, the circles are formed by contiguous boulders in
the manner of a kerb and on the basis of artefactual collections Herity
(1974) believes these sites should be assigned to the passage grave
tradition. More recent work suggests these sites predated the principal
passage grave construction period in the Boyne Valley by over 500 years
(Berenhult 1980).

The third group of sites is drawn from the Boyne Valley
passage grave cemetery located approximately 45 km north of Dublin.
Regrettably only 6 sites from the 16 surveyed sites were in any way
useful for analysis. Four of these sites are satellites of the main
Knowth tomb. The other sites are the large tomb of Dowth and the
smaller mound in the middle of the valley known as 'E'. The fourth
group of sites are the Wicklow-Kildare group spread along the western
edges of the Wicklow Mts. starting about 20 km southwest of Dublin and
running south for 30 km. Nine sites were surveyed and 5 are situated on
high hill tops and are almost certainly passage graves. The site on
Baltinglass Hill is known from excavations to have been built in at
least two phases and so these have been separated bringing the total
number of sites to 10. Two sites on the foothills, Athgraney and
Boleycarigeen, are stone circles and the last two sites, Broadleas and
Castleruddery, have an enigmatic architecture that draws on both the
passage grave and stone circle building traditions. Plans of all sites
can be found in Patrick (1978).

11   SIMULATIONS

While we have strong reasons to believe that the minimisation
of message length provides an absolute test for choice among competing
hypotheses, the theory underlying the method is as yet not widely known,
nor completely developed. Therefore it was considered desirable to
treat ΔI ($=I_p - I_T$) as just another test statistic, and to
investigate its sampling distributions under H_p and H_T. This study
is not yet complete but Monte-Carlo calculations have provided estimates
of sampling distributions of ΔI when H_T and H_p are restricted to
circular models, and the true population is either a population of Thom
circles, or a population of circles of arbitrary radius.

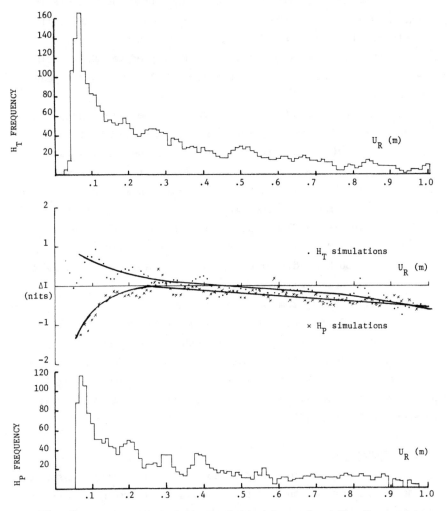

Fig. 3   Cork-Kerry simulations: Combined Groups I and II - Three point
moving average graph of ΔI as a function of U_R for H_P and H_T.
The histograms show the number of values used to determine each
average ΔI.

To ensure that these distributions would be relevant to the
field data, the artificial populations were designed to have the same
numbers and angular distributions of stones as the real sites, and to
have similar ranges of radii. As the real sites fall into two rather
distinct groups, two Monte Carlo calculations were made, one with
artificial populations resembling the Carrowmore sites, the other
mimicking the Cork-Kerry sites, which are rather smaller and have on
average only half the number of stones.

Figure 3 shows the results for the Carrowmore sites with 390
and 280 simulations of H_P and H_T circles respectively. This
figure is a plot of the mean ΔI as a function of the uncertainty in the
radius, U_R. The two curves in Figure 3 indicate that ΔI ceases to
distinguish H_P and H_T populations when U_R exceeds about .45
m. For very large U_R, ΔI always favours the H_P hypothesis
regardless of whether the population conforms to H_P or H_T,
because the quantisation of an H_T population is so swamped by noise
that there is no justification for ascribing any particular quantum
number to R. One can expect good discrimination between the hypotheses
if U_R is smaller than about .4 m but very little or no
discrimination for larger values of U_R. There were 720 and 1020
simulations of H_P and H_T respectively for the Cork-Kerry sites,
which yielded similar patterns except the limiting value of U_R was
about 0.30 m, which was caused by a smaller number of stones per site.

12   RESULTS

The discussions presented in this paper are confined to
looking at the overall and regional results. The details from each site
will be discussed elsewhere. All IMs were computed using the centroids
of the stones. There is some suggestion that more stones fall outwards
than inwards and the evidence for this is presented in Patrick (1978).

The implementation of the IM technique involves computing for
each site the I4 + I5 elements of the IM for each shape. Table 2 is a
list of these values for all the Carrowmore sites. It can be seen
readily that for many sites the IMs differ very little for the different
shapes. Take for example Carrowmore 19 where the difference between the
f_2, f_3 and f_{3S} IMs is only 0.57 nits. Any message that describes
a set of sites must communicate the dictionary of shapes available as
defined in the hypothesis statements and must describe the shape

Patrick & Wallace: Stone Circle Geometries

Table 2: The (I4 & I5) IM components of the Carrowmore sites for each shape. The a_{3A} and a_{3B} IMs refer to Thomsian Flattened circles but with the unbiased estimate of the radius rather than an integral MY radius.

| SITE | H_P | | | | | H_T | | | H_A | |
	f_0	f_2	f_3	f_{3s}	t_0	t_2	t_{3A}	t_{3B}	a_{3A}	a_{3B}
01	34.94	32.34	33.32	33.60	35.72	39.33	31.22	32.08	31.51	30.31
04	27.91	22.48	20.83	20.76	28.45	26.20	22.29	19.42	22.77	18.17
07	34.83	32.71	34.62	34.95	34.79	37.05	35.86	32.93	34.96	33.93
09	11.72	10.95	—	12.41	11.90	14.54	13.49	13.33	13.12	13.22
15	15.01	16.41	17.23	17.99	15.15	27.17	16.23	18.35	16.68	18.99
18	22.82	22.86	—	24.80	22.54	26.78	24.52	21.70	24.54	23.21
19	54.76	49.26	49.83	49.51	55.41	50.89	48.91	50.95	47.42	51.24
26	26.76	20.48	21.42	20.65	25.67	23.70	19.84	22.07	20.17	18.49
28	44.06	46.01	45.89	46.15	43.16	51.36	46.05	49.66	46.16	50.49
32	24.80	22.45	23.45	23.57	26.26	28.10	20.07	22.66	20.92	22.82
36	41.67	41.85	43.97	44.18	41.92	46.13	43.21	42.68	43.27	42.05
37B	25.93	25.35	24.37	26.01	26.75	31.50	27.70	27.44	26.80	26.68
48	10.74	11.30	—	13.37	9.84	13.74	12.57	12.71	12.94	12.80
57	34.92	29.08	27.09	26.64	35.60	34.40	27.72	31.68	28.71	30.21
TOTAL	410.87	383.53	—	394.59	413.16	450.89	389.68	397.66	389.97	392.61

allocated to each site. It is necessary to search throughout the table
of site IMs for the combination of site IMs and dictionary and shape
costs that minimise the total IM. An algorithm for a search technique
can be found in Boulton (1975).

Thom has argued that the British data on which he bases his
hypothesis should not be subdivided but studied as a total data set (see
Patrick and Butler 1974). To apply the IM technique in this manner one
must search for the optimum shape combination over the 65 Irish sites.
It was evident very early in the analysis that the shapes of f_3 and
t_2 (Thomsian ellipse) were of no use to their respective hypotheses.
The cumulative totals for each hypothesis for each data group are
presented in Table 3. The difference between the two hypotheses is 0.43
nits, which is insignificant. However it is evident that Thom's Type B
flattened circle is of no use and so need not be described in the
dictionary. The new total for H_T is 908.76 nits which gives H_T
a lead of 3.78 nits. This can be interpreted as an odds ratio of 44:1
($e^{3.78}$:1) in favour of H_T. This represents a marginal advantage
for H_T but must be viewed in the presence of the Il part of the IM,
where we believe Thom's hypothesis requires a much lengthier description
than does the Patrick hypothesis. A comparison of H_T amd H_P
solutions for Carrowmore 26 can be seen in Fig 4.

13   THE REGIONAL RESULTS

The overall results presented in the previous section ignore
any groupings of the data and merely provide the optimum IM on the
assumption that the shape frequencies are independent of any regional
groups. This is consistent with Thom's presentation of his own work on
English and Scottish data. However we consider that the Irish data has
a clearly defined classification of four groups based on geographical
distribution and archaeological evidence. There are many features of
the sites which differ markedly between different groups, yet are
relatively uniform within a given group (Herity, 1974; O'Nuallain,
1975). Were the data necessary to describe these features added to our
geometric data, there is little doubt that a message conveying the
enlarged data set would be minimised by a classification into groups
corresponding closely to the geographic groups. If our results show
geometric differences between the groups, this will reinforce the
validity of the classification.

Fig. 4 Comparative examples of H_T and H_P solutions for Carrowmore 26.

Broken line f_{3s} : $r_0 = 7.56$, $r_2 = -.43$, $r_3 = 0.22$, $\beta = 16^0$, $\sigma = .25$

Full line a_{3a} : $r = 7.82$, $\beta = 163^0$, $\sigma = 0.27$

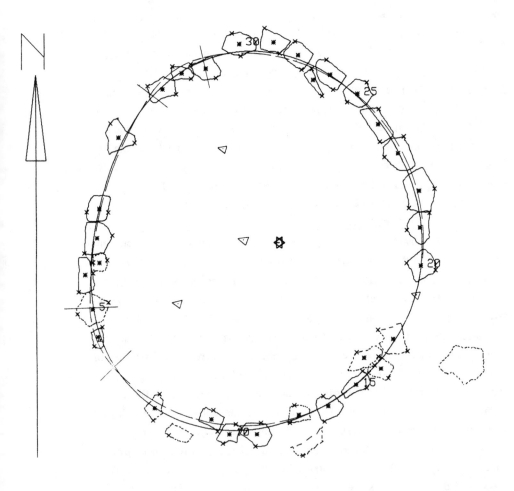

A message that is intended to describe the total group of 65 sites making use of their geographic grouping will now have to provide both class dictionaries and class labels that describe the allocation of each site to one of four archaeological groups. The sites are distributed in the proportions of 14:35:10:6 for Carrowmore, Cork-Kerry, Wicklow-Kildare and the Boyne Valley respectively. This has the cost of 80.45 nits but is the same for both H_P and H_T and so cannot contribute to discrimination between the two hypotheses. Now, for each group it is necessary to encode the dictionary of shape families available and the shape that each site belongs to. Table 4 presents a list of the IM for each regional group for a selection of shape

Table 3:   The total optimum IMs for each hypothesis over 65 sites. Dictionary and label costs are not included in each regional total but are only determined from the total distribution.

GROUP	H_P	Shape Frequency f_0, f_2, f_{3S}		H_T	Shape Frequency t_0, t_{3A}, t_{3B}
CARROWMORE	378.31	(7,6,1)		376.10	(8,6,0)
CORK-KERRY	276.56	(35,0,0)		270.99	(34,1,0)
WICKLOW	158.21	(3,5,2)		174.73	(5,5,0)
BOYNE VALLEY	48.33	(3,1,2)		51.25	(4,2,0)
	861.41			873.06	
DICTIONARY AND LABEL COSTS	51.13	(48,12,5)		39.05	(51,14,0)
	912.54			912.11	

Table 4: List of accumulated IMs plus shape label costs for each data group for a selection of the optimum shape combinations under H_p and H_T. Note that each table entry includes both a shape dictionary and shape label message cost according to the frequencies shown in brackets. Each group minimum is encircled.

SHAPES	CARROWMORE	CORK-KERRY	WICKLOW KILDARE	BOYNE VALLEY	TOTAL
H_p					
$f_{0,2,3,3S}$	393.80 (5,6,1,2)	295.84 (30,3,0,2)	171.40 (1,5,1,3)	58.23 (2,1,1,2)	919.27
$f_{0,2,3S}$	388.67 (0,12,2)	282.85 (35,0,0)	168.80 (1,6,3)	55.99 (4,0,2)	896.31
$f_{0,2}$	389.68 (5,9)	280.04 (35,0)	169.19 (2,8)	58.29 (3,3)	897.20
$f_{0,3S}$	392.20 (8,6)	281.54 (34,1)	173.30 (2,8)	54.45 (4,2)	901.49
$f_{2,3S}$	386.57 (12,2)	–	167.48 (7,3)	62.39 (4,2)	–
f_0	410.87	276.56	199.38	66.67	953.48
f_2	383.53	–	166.68	62.42	–
f_{3S}	394.59	–	172.34	–	–
H_T					
$t_{0,2,3A,3B}$	389.00 (6,0,5,3)	288.73 (32,0,2,1)	184.67 (3,3,1,3)	59.72 (3,0,2,1)	922.12
$t_{0,3A,3B}$	387.28 (6,5,3)	280.30 (35,0,0)	184.56 (5,5,0)	57.60 (4,2,0)	909.74
$t_{0,3A}$	386.79 (8,6)	277.46 (35,0)	182.70 (5,5)	56.02 (3,3)	902.97
$t_{0,3B}$	391.90 (6,8)	282.11 (33,2)	193.18 (5,5)	62.29 (3,3)	929.48
t_0	413.16	274.01	199.53	65.05	951.75
t_{3A}	390.53	–	181.12	62.46	–

combinations for both hypotheses. The number of sites assigned to each shape is shown in brackets. The results show that for both H_P and H_T the use of the complete range of shapes, i.e. using the absolute minimum IM for each site, is one of the most expensive ways of describing the data. The sixth column of totals shows that for H_P the optimum message is a $f_{0,2,3S}$ shape combination and for H_T it is $t_{0,3A}$. The detailed results show that for both H_P and H_T the Cork-Kerry group are best described as only circles with no ovoids at all. The Boyne Valley group has the minimum IMs for the $f_{0,3S}$ shapes under H_P and $t_{0,3A}$ under H_T. The Carrowmore and Wicklow groups yield the minimum IM under H_P as f_2 while under H_T the $t_{0,3A}$ shape combination is the minimum. Thus we can say that for H_P the f_0, f_2 and f_{3S} shapes are all needed to provide the optimum solutions and for H_T the t_0 and t_{3A} shapes only are necessary.

It is evident that most regional groups have substantially different optimal shape frequencies and so the cumulative totals produce a somewhat different result to the overall results presented previously. For H_P the $f_{0,2,3S}$ shape combination yields an IM of 896.31 nits whilst the H_T optimal shape combination is $t_{0,3A}$ with a value of 902.97. The difference represents an odds ratio of 780:1 (i.e. $e^{6.66}$: 1) in favour of H_P.

Under H_T the four groups yield t_0 to t_{3A} ratios of 8:6, 35:0, 5:5 and 3:3. Thus the only significant correlation of geometry with group is the absence of non-circular shapes in the Cork-Kerry group. Under H_P, the f_0, f_2, f_{3S} ratios are 0:12:2, 35:0:0, 1:6:3 and 4:0:2. This shows, besides the circularity of all Cork-Kerry sites, a significant difference between the fourth (Boyne Valley) group and all others. Thus H_P benefits more than H_T by a geographic classification.

It was requested, well before the development of this technique (Patrick 1975), that groups of sites already defined by prior research should be analysed independently. This request was based on the archaeological evidence that prehistoric Ireland and Britain was occupied by small independent tribal units. The different shape frequencies among regional groups revealed by this study are consistent with the spirit of the Patrick and Wallace hypothesis and vindicates the early considerations.

14 CONCLUSIONS

The presented results do not reveal a clear picture of which hypothesis is preferable. On a regional basis H_p is favoured at odds of 780:1. The regional group data indicate that H_p requires the three shapes f_0, f_2 and f_{3S} and that these three shapes have markedly different geographic distributions. This is gratifying as it is consistent with the general description of H_p. However it was unexpected that the f_0 shape would so completely dominate the Cork-Kerry group. In retrospect this is a sensible result as most of the sites have very few stones and so it is not possible to provide evidence to support the more complex designs. This problem introduced the question as to whether one should consider that only one shape family is present; i.e. f_{3S}. Then many sites would merely have zero fourier coefficients. Fourier family, f_{3S}, descriptions for all sites using zero coefficients would be more costly compared to f_0 or f_2 descriptions but would have no shape dictionary or label costs whatsoever. Whilst such a classification might be advantageous for describing substantial numbers of sites where archaeological classifications are ignored it would diminish the visibility of shape to region correlations that can appear, as, for example, in the Cork-Kerry group.

The Carrowmore and Wicklow results are gratifyingly consistent with H_p because for most sites it is not possible to discriminate between the use of f_2 and f_{3S} and to a lesser extent the f_0 shapes. Thus most sites have fairly ill-defined shapes which is consistent with our hypothesis. The comparison of the H_p results to H_T for Carrowmore indicates that Thom's flattened circle designs are quite efficient at describing these sites with their flattened facades. Thom's circle t_0 is an efficient shape though the simulations show there is no distinguishable difference between the two hypotheses for circles set out with the average uncertainty of these sites. The Wicklow sites provide an identical picture for H_p where f_0, f_2 and f_{3S} are highly competitive with each other. However Thom's hypothesis is hopelessly inefficient at describing these sites.

The Cork-Kerry results provide a direct comparison of just the circle shapes without the complexities of shape labels. The slightly shorter message length for H_T suggests there may be some

merit in Thom's arguments for a quantum. However, the advantage held
by H_T is not as large as one would expect from the simulation results
were H_T true.

We have investigated a weak pacing hypothesis, viz. that each
circle was set out with a radius equal to an integral number of paces,
where the "pace" for each site is a random variable drawn from
$N(0.8,0.1)$ (in metres). This hypothesis leads to a prior distribution
of radii having modest peaks up to about 5 m, but thereafter virtually
uniform. Its use leads to an IM (I4 + I5) for the Cork-Kerry sites
smaller than does H_T. Thus, the advantage of H_T over H_P for
these sites is evidence only of a very rough quantisation.

If we restrict ourselves to just the circles f_0 and
t_0 we can say for both the Carrowmore and Cork-Kerry sites that
H_P and H_T generate essentially the same message length. However,
H_P describes the sites on the assumption that the radii are uniformly
distributed, but to achieve this description an optimal quantal
subdivision of the data is determined. The nodal positions of this
subdivision are separated at the uncertainty in the radius, U_R,
which averages .42m and .28m for eight Carrowmore sites and the
Cork-Kerry group respectively. This gives average nodal points at .42
and .28m intervals respectively, which, averaged over the 43 sites, is
.30. Now Thom's hypothesis as we infer it from his published results is
that 1 my radii, i.e. .829 m occur with a frequency of .7 and $\frac{1}{2}$ my
radii, i.e. .415 m, have frequency .3. Thus if one conjures up a single
"effective" quantum that would occur with equal frequency it would seem
to be a value between .415 and .829 m. The conclusion is that because
of the inaccuracies of the data, a description of sites having uniformly
distributed radii using Thom's t_0 shapes, is little different to
describing the sites as having uniformly distributed radii with an
average uncertainty of measurement of about 0.3 m. Thus Thom's quantum
hypothesis imitates very closely the random hypothesis in the area of
uncertainty that the Irish data falls into.

The evolution of Thom's hypothesis can now be seen in a new
perspective. His initial hypothesis claimed the existence of a quantum
of 5.44 ft (Thom, 1955). Then the quantum was revised to 2.72 ft
(Thom, 1961). This was followed by the addition of the 1.36 ft quantum
at a frequency of 30% (Thom, 1967). The last of the inconsistent
observations that were not satisfactorily explained were subjected to

perimeter conditions (Thom 1967). Thus we can see that as more data became available a deepening complexity evolved to successively pick up the inconsistent observations. There can be little doubt that a "megalithic yard" was never used to set out the Irish sites. Thom's flattened circle geometric design is a sensible interpretation of the stone positions but certainly fits no better and often much worse than the fourier circles. Thom's hypothesis is so complex and involved that it takes many pages of written text to describe. The Patrick and Wallace hypothesis on the other hand can be presented much more briefly and must be favoured a priori. Given this condition H_T must be rejected in favour of H_P for the Irish data analysed herein.

The two previous analyses of Thom's stone circle diameters have considered that only a 1 my quantum was used (Kendall 1974; Freeman 1976). It is possible to use the IM results to remove the effect of the 1/2my quantum. There are 23 sites that use a 1/2my quantum for the optimum solution and likewise 42 sites that use a 1my quantum. These 23 sites must be changed to the nearest 1my quantum which results in adding 67.21 nits to the overall total. However as the probability of 1my quantum is now 1 so there is a total correction to the 65 sites of -42.67 nits. Thus the overall result of removing any 1/2my solutions is to add 24.54 nits to I_T. Whilst this is an over-estimate of the correction, because changes in the optimum shape have not been taken into account, it is evident that the simplified H_T would fare even worse than it does at present against H_P.

The group results demonstrate the importance of considering the sites in their archaeological groupings. The Cork-Kerry stone circles are circles some of which are set accurately and some rather carelessly. The design plans of the passage graves are certainly not circles but are the product of other architectural considerations where the builders created a wide somewhat flattened facade to give an impressive entrance to the tomb. The kerb was completed to present a continuous wall and positioned to look like a neat curve without perturbations that would offend the eye whilst maintaining the general scale of the monument as defined by the size of the front facade region. As each small section of the kerb was kept smooth, large scale variations were imperceptible, resulting in the kerb having fluctuating sharper and flatter curves that are only now made apparent by our modern-day plans.

Professor Thom's geometric designs and megalithic yard are, in our opinion, somewhat extravagent extrapolations of the evidence available. His hypothesis is not competitive for the Irish sites tested herein and we feel this must intimate a similar result for the British sites once they are evaluated by the technique developed here.

BIBLIOGRAPHY

Angell, I.O. (1976). Stone circles: megalithic mathematics or neolithic nonsense? Math.Gaz., 60, 189-93.
    - (1977) Are stone circles circles? Science and Arch. no.19, 16-9.
Barber, J. (1972). The Stone Circles of Cork and Kerry: A Study. M.A Thesis. National University of Ireland, University College, Cork.
    - (1973). The orientation of the recumbent-stone circles of the southwest of Ireland. J. Kerry Hist. Arch. Soc., 6, 26-39.
Berenhult, G. (1980). The Archaelogical Excavation at Carrowmore, Co. Sligo, Ireland, Excavation Seasons 1977-79. Stockholm: Institute of Archaeology.
Boulton, D.M. (1975). The Information Measure Criterion for Intrinsic Classification. Ph.D. Thesis, Monash University, Australia.
Buckley, J.W., Buckley, M.H.& Hung-Fu Chiang (1976). Research Methodology & Business Decision. New York: Nat.Assoc.Accts.
Burl, H.A.W. (1976). Stone Circles of the British Isles. New Haven: Yale University Press.
Burgess, C. (1974). The Bronze Age. In British Prehistory: A New Outline, ed. C. Renfrew, pp223-32. London: Duckworth.
Cowan, T.M. (1970). Megalithic rings: their design construction. Science, 168, 321-5.
Emory, C.W. (1980). Business Research Methods. New York: Richard D. Irwin.
Freeman, P. (1976). A Bayesian Analysis of the Megalithic Yard. Jnl.R.Stat.Soc.,A, 139(1), 20-55.
Halstead, M.H. (1977). Elements of Software Science. New York: Elsevier North-Holland.
Herity, M. (1974). Irish Passage Graves. Dublin: Irish University Press.
Kendall, D.G. (1974). Hunting Quanta. Phil. Trans. R. Soc. Lond., A.276, 231-266.
O'Nuallain, S. (1975). The Stone Circle Complex of Cork and Kerry. Jnl. Royal Soc. Antiq. Ireland, 105, 83-131.
Patrick, J. (1975). A reassessment of Megalithic Astronomy and Geometry, Conference on Ceremonial Science and Society in Prehistoric Britain, Glasgow.
    - (1978). An Information Measure Comparative Analysis of Megalithic Geometries. Ph.D. Thesis, Monash University, Australia.
Patrick, J. and Butler, C.J. (1974). On the interpretation of the Carnac menhirs and alignments by A. and A.S. Thom including Thom and Thom's reply. Irish Arch. Res. Forum, 1, no.2, 29-44.
Solomonoff, R.J. (1964). A formal theory of inductive inference. Info.& Con., 7, 1-22, 224-58.
Thom, A. (1955). A statistical examination of the megalithic sites in Britain, Jnl. R. Stat. Soc., A. 118, no. 3, 275-95.

- (1961). The egg shaped standing stone rings of Britain.
Archives Internationales D'histoire Des Sciences, $\underline{14}$, 291–303.
-(1967). Megalithic Sites in Britain. London: Oxford
University Press.

Wallace, C.S. and Boulton, D.M. (1968). An information measure for
classification. Computer J. $\underline{11}$, no. 2, 185–94.

Wallace, C.S. and Boulton, D.M. (1975). An Invariant Bayes Method for
Point Estimation. Class. Soc. Bull., $\underline{3}$, no.3, 11–34.

THE PRESENT POSITION OF ARCHAEO-ASTRONOMY

OLAF PEDERSEN

Aarhus University Denmark and Cambridge University England

An international conference on 'archaeo-astronomy' is an interesting experience for a historian of astronomy of the traditional school. Even if
he is somewhat acquainted with the literature on the subject, and perhaps to some
extent has also used it for teaching purposes, he must feel very much an outsider.
He is familiar only with certain types of sources in the form of documents or
instruments and not used to the strange array of material presented here in evidence of a startling hypothesis, of a prehistoric prelude to an historical account
which until now has begun with the earliest written records. The immediate
effect is a feeling of perplexity or confusion which only gradually gives way to a
set of more structured impressions. Certain parts of the material seem to
indicate possible fields in which the possibility of fruitful research cannot be dismissed without further investigation. A number of more or less obvious pitfalls
reveal themselves as almost inevitable consequences of the very nature of the
material evidence. Furthermore, the dangers connected with any kind of inter-
disciplinary study become clearly apparent. And last, but not least, one becomes increasingly aware of a series of methodological questions which must be
answered if the whole subject is to be salvaged from the rocks, lurking under the
surface in the form of arbitrary assumptions, vicious circles, and unjustified inferences. However, the following pages do not aim at any systematic or complete
examination of all such problems. Their only purpose is to register, in a rather
personal way, some of the more obvious features of the astro-archaeological
debate which make one pause and reflect.

Thus one cannot help being struck by the wide range in space and
time of the material offered here for the consideration of 'archaeo-astronomers'.

At the one end of the scale there are Amerindian data which seem to be of fairly recent origin. Some of them may even be connected with the remnants of a living tradition which may be picked up by linguists or ethnographers working in the field - a fascinating perspective which makes the traditional historian realise the inadequacy of his customary methodological equipment. From the New World come also the documents and inscriptions produced by Mezo-American literate societies less than a thousand years old. Their deciphering and interpretation has already established their calendaric and astronomical importance beyond any doubt, reminding the historian of the not radically different task begun a hundred years ago when Strassmaier, Kugler and Epping first set to work on the Mesopotamian cuneiform clay tablets in the British Museum. Consequently, he will feel more at home in this field and confident of the methods employed by the brave handful of experts and specialised scholars whose achievements the rest of us have to observe with distant awe and admiration. Finally, at the other end of the scale, we are confronted with all the 'megalithic' material from the Old World in the form of stone monuments set up by illiterate societies four or five thousand years ago and giving rise to methodological problems and questions of interpretation of a different nature to which we shall return in a moment.

Thus there are many abodes in the realm of 'archaeo-astronomy' and it is perhaps not entirely irrelevant to suggest that the confines of this very concept are not yet clearly delineated. The term as such points to an account of some kind of astronomical activity of the past based on material brought to light through archaeological research. But if this were taken as a definition of the concept we should have to exclude everything based on the written books of the Middle American cultures which play such a conspicuous role in the 'archaeo-astronomy' of the New World. Moreover, we should also have to include the astronomy of the ancient Middle East the knowledge of which is exclusively based on tablets excavated by archaeologists, but usually disregarded by 'archaeo-astronomical' conferences or journals. Other such inconsistencies could be mentioned; but perhaps they should not be taken too seriously at the present stage. Some terminological confusion will usually accompany the establishment of a new subject. But only time can show whether the concept of 'archaeo-astronomy' can be clarified to such a degree that a well defined new area of historical re- search can be said to have emerged and gained universal recognition.

Turning now to the more restricted field of 'megalithic' astronomy, the lack of terminological precision seems less marked, at least at first sight, since the primary material offered in evidence is of a more uniform and undoubt- edly archaeological character, although by no means without diversity as even a brief glance at some of the recent publications will show: small or large 'monu- ments' of natural or dressed stones set up in the field in more or less regular geometrical patterns, similar structures destroyed long ago and now only dis- cernible from half obliterated stone-sockets or post-holes - but also artificial floors of pebbles on a ledge of a hillside, or natural notches in distant mountain ridges across the sea - all of it things with which the archaeologist is familiar and which the surveyor can describe in precise geometrical and numerical terms. That the dating and building history of these remains must be of prime importance to the study of European prehistory has long been self-evident, and much work has been done in order to reveal their original function or significance. No absolute and undisputed solution has been found, and Joshua's old question: What Mean These Stones? still needs a decisive answer. In this situation a historian of astronomy of the old school must naturally prick up his ears when he is told that astronomy may provide at least a part of the answer to the question, or rather to two separate questions, viz.

(1) Can astronomy contribute to explaining the geometry of mega- lithic monuments as carefully established by good surveying, and

(2) Can the geometrical structures of these monuments (or some of them) be admitted as evidence of any true insight in astronomy on the part of their builders?

These two questions are, of course, closely connected. Neverthe- less, it is perhaps helpful to keep in mind that they are separate, and that it might be dangerous not to distinguish them as sharply as possible.

The first to deal seriously with these questions was Sir J. Norman Lockyer whom we must ever regard as the founding father of 'megalithic astronomy'. As is well known Lockyer began by determining the orientations of Egyptian pyramids and temples, trying to correlate their main axes of symmetry with the azimuth of the rising Sun, or Sirius, on important dates of the ancient Egyptian year. He then went on with some much more accurate field work at Stonehenge and a number of other megalithic monuments in the United Kingdom at each of which he tried to prove the astronomical significance of a number of 'lines' more or less clearly marked by the geometrical structure of the monument in question. Lockyer showed how a simple formula in spherical astronomy, in connection with the formula giving the secular variation of the obliquity of the ecliptic (taking account also of terrestrial refraction and the elevation or depression of the visible horizon), could provide the date when the monument was built - under the fundamental assumption that a characteristic structural line or direction was meant to point to that point on the horizon where the sun or a certain star would rise on a particular date of the year. He crowned his efforts by calculating that if the axis of symmetry of Stonehenge was meant to indicate the rising sun on Midsummer Day's morning, the monument must have been built around 1680 B. C.- in agreement with the date assigned to it by the archaeologists of his time.

This apparently brilliant result aroused much popular interest and very little scholarly opposition. Soon after, Lockyer had a number of followers both in Britain and on the Continent who subjected an increasing number of mega- lithic sites to astronomical analysis. Thus it would seem that the implication of Lockyer's achievement was that the first of the two questions above must be an- swered in the affirmative: astronomy had proved itself to be a valuable tool for the interpretation of at least some of these monuments.

However, looking back upon this episode we have to be a little more circumspect. On the one hand it must be acknowledged that Lockyer's method was sound, viewed from a purely astronomical point of view, and that it is still, with further refinements,   the basis of all attempts to date structural lines in ancient monuments by astronomical means. On the other hand it must also be ad- mitted that there were arbitrary elements in Lockyer's work. For example, he stated no general criteria for the selection of such lines but simply picked those which appealed to him as the most promising, a fact which in the long run contributed to make more critical scholars suspicious of the method as such. Here it is sufficient to remember the fact that Lockyer first carefully determined the azimuth of the best axis of symmetry of Stonehenge, and next abandoned this line in favour of a slightly different direction. This new line passed through Stonehenge and three other ancient monuments, and Lockyer was unable to resist the temptation to consider all four sites as parts of a single structure, twenty miles long and assumed to have been originally devised for astronomical purposes. When in the 1950s Professor Atkinson pointed out that two of the four sites were more than a thousand years younger than Stonehenge, the reason for connecting them disappeared and Lockyer's adjusted axis of Stonehenge was exposed as an arbitrary and fictive construction of his imagination. This episode is a good illustration of one of the dangers of interdisciplinary projects: Lockyer was an expert in astronomy, but an amateur in archaeology, and his competence in the one field was not automatically transferred to the other.

In the meantime 'megalithic astronomy' had become suspect for very different reasons. Between the two World Wars it was eagerly cultivated in Germany by a group of publicists led by O. Reuter who wrote a vast, but un-critical and unmethodical compilation with the revealing title "Germanische Himmelskunde". The idea was that the real or alleged astronomical orientations of ancient monuments marked the beginning of astronomy and testified to the scientific and intellectual superiority of the 'Aryan Race' on German soil. Thus 'archaeo-astronomy' was exploited as an element of the ideological preparations of World War II and it was only too easy to dismiss it as just another crazy in-stance of Nazi propaganda. Fortunately this was a passing episode. But it is perhaps not amiss to remember it as a reminder of the fact that a scientific hypothesis can (given the right circumstances) appeal so strongly to wider circles that it becomes a popular myth serving non-scientific purposes. Of course, today 'archaeo-astronomy' does not have any mythical function - or does it?

Thus the 'archaeo-astronomical' hypothesis seemed to be discredi-ted in more ways than one precisely at the time when Professor Alexander Thom quietly began that patient and meticulous work which once more would transform it into a scholarly and respectable discipline. It would be presumptuous here to go into detail with this work, but it is only proper to mention some of the general features which are immediately recognisable even by an outsider.

First one cannot but admire the uncompromising discipline of Professor Thom's field work which has resulted in measurements and surveys of a technical excellence never reached before.

Secondly, adopting Lockyer's general method, Alexander Thom has refined and improved it by a number of special investigations of very great value also in other contexts, such as his detailed empirical study of terrestrial refraction and its dependence on the temperature and other conditions of the atmosphere.

Thirdly, Alexander Thom has consciously tried to reduce the arbitrary elements of Lockyer's approach by measuring hundreds of sites, thereby providing primary data in such numbers that statistical methods can be brought into play.

Such spectacular and indisputable achievements seem to indicate that at long last 'archaeo-astronomy' has come of age. It would indeed be unwise to deny that it now rests upon a more solid and extended basis of data than ever before. But, on the other hand, it would also be unwise to deny that the splendour of the new array of data has had a kind of blinding effect, obscuring the fact that the problem of how to interpret these data seems no closer to an universally acceptable solution than in December 1972, when scholars first met to discuss it at the joint meeting of the Royal Society of London and the British Academy. During the intervening years, one can even observe a kind of increasing polarisation between two groups. One of these gives very wide support to the general 'archaeo-astronomical' hypothesis, answering 'yes' to both the fundamental questions raised above. The other adopts a more reserved attitude and seems to be less and less inclined seriously to consider any general claim made on behalf of the hypothesis. It is certainly not the task of an outsider to judge between such positions each of which must be assumed to rest upon serious considerations by competent workers in this difficult area. But even an outsider without any archaeological work to his credit might venture a few remarks on the possible reasons for this general lack of consensus. I shall make a few observations such as may contribute to explain the fact that some scholars are still uneasy about the claims made by the more convinced adherents of the hypothesis.

Firstly it is important to remember that we meet with a rather complex set of claims. Much of Alexander Thom's earlier work was concerned with prehistoric metrology and geometry in Britain. This led to his assumption of the existence of a unit of length called the 'megalithic yard', the length of which (about 82 cm) was derived from statistical considerations of measurements of a very large number of stone 'circles'. This went hand in hand with the establishment of a typology of these monuments, based on geometrical interpretations of their various lay-outs, and enabled him to determine the nature and range of the knowledge of geometry possessed by the architects who planned them. This line of research was an ingenious exercise in prehistoric mathematics and completely independent of any considerations of prehistoric astronomy. Those who were present at the meeting in 1972 in London will remember the incipient enthusiasm, almost amounting to elation, in the audience when Professor Kendall presented his calculation, based on a new and highly sophisticated statistical method, of the significance level of the 'megalithic yard', proving that the probability of the measured diameters being quantified by chance was less than one per cent. But some of the participants may also have observed that this important result in the field of ancient metrology in some mysterious way seemed to lend support also to the astronomical hypothesis discussed at the same meeting. Credibility was so to speak transferred from one area to the other. Since the two areas are unconnected this was a logical mistake, possibly due to some psychological mechanism operating below the conscious level. The consequence was that the credibility of the astronomical claims seemed to be diminished as soon as the logical error of the transfer dawned upon us.

The second point is concerned with the almost inevitable dangers inherent in any kind of interdisciplinary research where the specialist becomes an amateur as soon as he ventures outside his own field. Perhaps this danger is greater for those born on the astronomical side of the blanket than for their more historically trained colleagues. 'Astronomers' may well be more tempted to

venture into an area dominated by the ordinary language of archaeology and history, than historians or archaeologists are tempted to move into a field where all
serious discourse presupposes the mathematical language of astronomy or statistics. This danger seems to be clearly apparent in much of the talk of the so-
called 'Megalithic Man'. Which status can be ascribed to this enigmatic crea-
ture? Was he a member of a particular race carrying megalithic architecture
from one coastal area of Europe to another, and possessed of particular astronomical insights peculiar to his tribe or society? Or was he simply a convenient
term for Man in any society at that stage in history at which he had learnt how to
master the technology of building in large stones? It seems to me that the protagonists of the hypothesis of 'Megalithic Man' as an astronomer must engage in
much more serious discussion with the experts in prehistory and archaeology, if
the validity of this concept is to be clearly established.

The same conclusion can be reached by way of a different argument.
A scholar trained in natural science usually operates with a clear criterion for
accepting an idea as a working hypothesis, viz. that it fits all, or most, of the
data at his disposal. This has proved to be a reasonable procedure when we are
concerned with data derived from nature. But it is not obvious that this procedure is equally justified with regard to data derived from artefacts or monuments
made by man in specific social conditions. Must we not here apply two different
criteria, viz.

I the hypothesis is acceptable if it fits all, or most of the known
data, and

II if it agrees with the intellectual and technical level of the society
which produced the remains from which the data are derived?

Of course, the second criterion means that a hypothesis based on the first criterion must be submitted to the judgment of scholars who study the intellectual or
technological level of past societies, and evaluated in the light of what they have
been able to establish from independent considerations of the source material.
Here the one unforgivable sin is to pretend that the intellectual level mentioned in
the second criterion can be established by means of the first without additional
support. This would be a vicious circle; but has it always been avoided? In

other words, does the hypothesis of a 'megalithic astronomy' entitle us to popu-
late neolithic Britain with a society of astronomer-priests when this idea is con-
tradicted by practically everything we know about megalithic society from other
sources? Or, to take only one further example, does the existence of a megalithic unit of length used up and down the British Isles and in Brittany entitle us
to assume the existence of a central megalithic 'Bureau of Standards' distributing
rods of standard length to local communities far and near, particularly when this
unit is so close to the length of an ordinary pace? These are not rhetorical questions, but problems which have to be carefully discussed before the hypothesis of
'megalithic astronomy' or 'megalithic mathematics' can make any serious impact
upon our notions of prehistory.

In this connection a word must be said about the problem of accuracy. Scientists have normally a great respect for accuracy; but perhaps this
can be exaggerated. It is quite possible that the quest for accuracy may develop
into a kind of obsession which leads to eccentric results. It is one thing to verify
that a 'line' can be related to a certain azimuth with a definite degree of accuracy.
It seems to be a very different thing to assume that this degree of accuracy was
intended by the architect who first laid out the line. At least some of the more
sophisticated interpretations of lines supposedly connected with the finer details
of the motion of the moon - interpolation procedures performed at some of the
fan-shaped monuments - seem to be a serious impediment to the application of
the second criterion above.

Finally one has to raise the problem of the selection of the material
upon which the hypothesis is founded. As already mentioned, this material is
now very extensive and derived from measurements of hundreds of monuments
each of which shows one or more 'lines' of possibly astronomical significance.
The existence of such lines seems established beyond doubt as seen from Professor Thom's histograms of declinations. Nevertheless, there remains a slight
feeling of uneasiness when one remembers that the number of monuments providing such lines seems to be very small compared with the total number of monuments which exist or is known to have existed. How would it affect our statistical
considerations if all the monuments were taken into account? Statistics is, as we

all know, not only the science of calculating probabilities; it is also the art of asking the right kind of questions on a data-base adequate to the provision of a reply. Not being a statistician, I wonder whether the following are legitimate questions: given a set of m monuments containing a total number l of established astronomical lines; what would be the number $N(m, l)$ of monuments which would have to be built at random in order to produce the given set? And how would it influence our attitude towards the hypothesis of a 'megalithic astronomy', if it should so happen that $N(m, l)$ proved to be of the same order of magnitude as the number of actual monuments? Perhaps this is an impossible problem, although a similar problem has already been solved in connection with one of the famous sites in Brittany; it is only stated here in order to suggest that the hypothesis cannot be adequately supported simply by calculating some kind of probability that merely affirms the established identity of some of the 'astronomical' monuments.

At the end it seems that we are left with many open questions and few definite answers. On the one hand it is impossible to accept all the claims made for the case for 'megalithic astronomy', for a number of reasons some of which have been suggested above. On the other hand, there are also too many solidly established results to justify its being summarily dismissed. So the case remains open, and the hearing must go on. Perhaps we are entering a phase where continuing research should concentrate less on the provision of more numerical material than on a critical analysis of the nature of the arguments used for purposes of interpretation, in order to disclose those danger-points where we may have been led astray by unwarranted assumptions or untenable conclusions.

INDEX